INTRODUCTION TO EUROCODE 2

Design of concrete structures (including seismic actions)

Derrick Beckett

Visiting Fellow in Structural Design, University of Greenwich

and

Andrew Alexandrou

Formerly Principal Lecturer, School of Civil Engineering, University of Greenwich

E & FN SPON

An Imprint of Chapman & Hall

London · Weinheim · New York · Tokyo · Melbourne · Madras

Published by E & FN Spon, an imprint of
Chapman & Hall, 2–6 Boundary Row, London SE1 8HN, UK

Chapman & Hall, 2–6 Boundary Row, London SE1 8HN, UK

Chapman & Hall GmbH, Pappelallee 3, 69469 Weinheim, Germany

Chapman & Hall USA, 115 Fifth Avenue, New York, NY 10003, USA

Chapman & Hall Japan, ITP-Japan, Kyowa Building, 3F, 2–2–1 Hirakawacho, Chiyoda-ku, Tokyo 102, Japan

Chapman & Hall Australia, 102 Dodds Street, South Melbourne, Victoria 3205, Australia

Chapman & Hall India, R. Seshadri, 32 Second Main Road, CIT East, Madras 600 035, India

First edition 1997

© 1997 D. Beckett and A. Alexandrou

Typeset by Florencetype Ltd, Stoodleigh, Devon
Printed in Great Britain by TJ International Ltd, Padstow

ISBN 0 419 20140 8

∞ Printed on acid-free text paper, manufactured in accordance with ANSI/NISO Z39.48–1992 (Permanence of Paper).

CONTENTS

PREFACE

The Faculty of the Built Environment, University of Greenwich, is committed to introducing a European dimension to its undergraduate and post-graduate teaching programmes. This includes language skills, social and economic issues, management and construction technology. The structural Eurocodes form part of the construction technology programme, and the purpose of this book is to introduce built environment students and graduates to the application of Eurocode 2 – Design of concrete structures, Part 1: General rules and rules for buildings, DD ENV 1992-1-1: 1992 (hereinafter referred to as EC2) to the design of conventional reinforced concrete buildings. The contents of the book are based on material presented at lecture courses and seminars held in the United Kingdom, Greece and Cyprus. As six of the European Community member states have national seismic design regulations, it was considered appropriate to include material on seismic actions and structural response, and thus reference is made to Eurocode 8 (Draft) – Design provisions for earthquake resistance of structures.

As the emphasis of the book is on applications, a complete chapter is devoted to the design of elements of a multi-storey reinforced concrete framework including seismic actions. In order to make the book as concise as possible, comprehensive appendices includes guidelines for preliminary design, design charts, data sheets and comparisons with BS 8110: 1985. Throughout the text, reference is made to the National Application Document (NAD) for use in the United Kingdom with ENV 1992-1-1: 1992. The book will also be of interest to users of the recently published Seismic Code for Reinforced Concrete Structures in Cyprus.

The Structural Eurocodes are in a state of continuous development and reference should be made to the latest issue of *Euronews Construction*, published for the Department of the Environment by *Building*. In the August 1993 issue there is a comprehensive review of the status of the Structural Eurocodes. Copies may be obtained from DOE, 2 Marsham Street, London SW1P 3EB (Tel: 0171 276 6596).

A 'Concise Eurocode' for the design of concrete buildings has been published by the British Cement Association (BCA) and copies may be obtained from the BCA, Century House, Telford Avenue, Crowthorne, Berkshire RG11 6YS.

A disc with all the software listed in the text is available from the authors, who can be contacted through the Faculty of the Built Environment Business Centre.

In the interval between the completion of this text and its publication, there has been continuous development in the drafting of Eurocodes and the current status (May 1997) for Eurocodes 1 and 2 is as below:

Eurocode 1: Basis of design and actions on structures

		UK draft available	*Euronorm due*
Part 1:	Basis of design	Sept 96	Mid 1999
Part 2-1:	Actions on structures Densities, self-weight and imposed loads	Oct 96	Mid 1999
Part 2-2:	Actions on structures Actions on structures exposed to fire	Sept 96	Mid 2000
Part 2-3:	Actions on structures Snow loads	Apr 96	Mid 1999
Part 2-4:	Actions on structures Wind actions	Nov 96	Mid 1999
Part 2-5:	Actions on structures Thermal actions	Sept 97	
Part 2-6:	Actions on structures Construction loads and deformations imposed during construction	Jan 98	
Part 2-7:	Actions on structures Accidental actions	Sept 97	
Part 2-xx:	Actions on structures Actions from currents and waves	Postponed	
Part 3:	Traffic loads on bridges	Apr 97	Mid 2000
Part 4:	Actions in silos and tanks	Jun 96	2001
Part 5:	Actions induced by cranes and machinery	Jan 98	

Eurocode 2: Design of concrete structures

	UK draft available	Euronorm due
Part 1-1: General rules		
General rules and rules		
for buildings	May 92	Late 2000
Part 1-2: General rules		
Structural fire design	Jul 96	Mid 2000
Part 1-3: General rules		
Precast concrete elements		
and structures	Sept 96	Mid 1999
Part 1-4: General rules		
Structural lightweight		
aggregate concrete	Sept 96	Mid 1999
Part 1-5: General rules		
Unbonded and external		
tendons in buildings	Sept 96	Mid 1999
Part 1-6: General rules		
Plain concrete structures	Sept 96	Mid 1999
Part 2: Reinforced and		
prestressed concrete		
bridges	Sept 97	
Part 3: Concrete foundations	Jan 98	
Part 4: Liquid retaining and		
containment		
structures	Jan 98	
Part 5: Marine and maritime		
structures	Postponed	
Part 6: Massive structures	Postponed	

Derrick Beckett and Andrew Alexandrou
June 1997

ACKNOWLEDGEMENTS

This book is based on lecture material presented at the University of Greenwich and elsewhere, and the authors are indebted to David Wills, Dean of the Faculty of the Built Environment, and Lewis Anderson, Deputy Head, School of Land & Construction Management, for giving them the opportunity to write this book during the spring and summer of 1993. Thanks are due to Sue Lee for preparing the diagrams for Chapters 2–5 and 8 and Appendices B–E and to Jenny Lynch for assistance with production.

Extracts from British Standards are reproduced with the kind permission of BSI. The extracts are as follows and are also indicated where they occur in the text.

DD ENV 1992-1-1: 1992, Eurocode 2 – Design of concrete structures, Part 1: General rules and rules for building (together with United Kingdom National Application Document): clause 2.1 – P(1) to P(4); table 2.2; table 4.1; table 4.2; clause 2.5.1.2 – P(1) to P(5); figures 2.2, 2.3 and 2.4; clause 2.5.3.3 – P(1) to P(6); figures 4.3 and 4.5; clause 3.2.4.2 – P(2); figure 4.12; figure 4.15; figures 4.18, 4.19 and 4.20; clauses 4.4.1.1, 4.4.2.1 and 4.4.2.2; tables 4.11, 4.12, 4.13 and 4.14; clauses 4.4.3.1 and 4.4.3.2; figures 5.1 and 5.2; table 1 (NAD) and table 5 (NAD)

ENV 206: 1990: table 3

BS 8110: 1985: Structural use of concrete, Part I: tables 3.2 and 3.4

Complete copies can be obtained by post from BSI Sales, Linford Wood, Milton Keynes MK14 6LE.

Special thanks are due to Sally Beckett, who was responsible for typing the text and the general coordination of the book.

NOTATION

The principal symbols used in the text are listed below and others are defined within each chapter.

Latin upper-case symbols

A_c Total cross-sectional area of a concrete section

A_s Area of reinforcement within the tension zone

A_{sw} Cross-sectional area of shear reinforcement

E_{cm} Secant modulus of elasticity of normal weight concrete

E_s Modulus of elasticity of reinforcement

M_{sd} Design value of the applied internal bending moment

N_{sd} Design value of the applied axial force (tension or compression)

T_{sd} Design value of the applied torsional moment

V_{sd} Design value of the applied shear force at the ultimate limit state

Latin lower-case symbols

$1/r$ Curvature at a particular section

b Overall width of a cross-section, or actual flange width in a T or L beam

d Effective depth of a cross-section

b_w Width of the web on T, I or L beams

f_c Compressive strength of concrete

f_{cd} Design value of concrete cylinder compressive strength

f_{ck} Characteristic compressive cylinder strength of concrete at 28 days

f_{cm} Mean value of concrete cylinder compressive strength

f_{ctk} Characteristic axial tensile strength of concrete

f_{ctm} Mean value of axial tensile strength of concrete

f_y Yield strength of reinforcement

f_{yd} Design yield strength of reinforcement

f_{yk} Characteristic yield strength of reinforcement

f_{ywd} Design yield strength of stirrups

h Overall depth of a cross-section

l Length; span

l_{eff} Effective span of a beam

s Spacing of stirrups

u Perimeter of concrete cross-section, having area A_c

x Neutral axis depth

z Lever arm of internal forces

Greek symbols

γ_c Partial safety factors for concrete material properties

γ_G Partial safety factors for permanent actions G

γ_Q Partial safety factors for variable actions Q

γ_s Partial safety factors for the properties of reinforcement

ε_u Elongation of reinforcement at maximum load

ε_{uk} Characteristic uniform elongation of reinforcement at maximum load

ρ_l Reinforcement ratio for longitudinal reinforcement

ρ_w Reinforcement ratio for shear reinforcement

σ_c Compressive stress in the concrete

φ Diameter of a reinforcing bar

ψ_0 Combination factor for rare load combinations

ψ_1 Combination factor for frequent load combinations

ψ_2 Combination factor for quasi-permanent load combinations

Other symbols

These are defined separately within the text.

THE EUROPEAN COMMUNITY

Key:

1. **Austria** (Capital – Vienna)
2. **Belgium** (Capital – Brussels)
3. **Denmark** (Capital – Copenhagen)
4. **Eire** (Capital – Dublin)
5. **Finland** (Capital – Helsinki)
6. **France** (Capital – Paris)
7. **Germany** (Capital – Berlin)
8. **Greece** (Capital – Athens)
9. **Italy** (Capital – Rome)
10. **Luxembourg** (Capital – Luxembourg)
11. **The Netherlands** (Capital – Amsterdam)
12. **Portugal** (Capital – Lisbon)
13. **Spain** (Capital – Madrid)
14. **Sweden** (Capital – Stockholm)
15. **United Kingdom** (Capital – London)

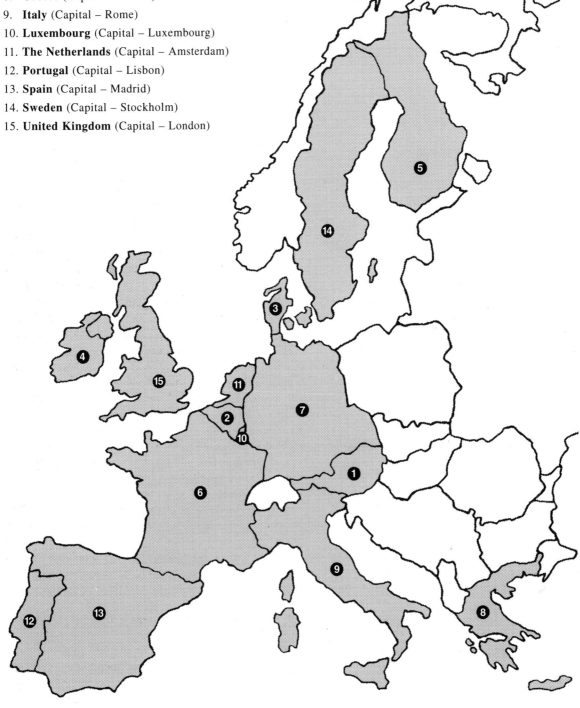

1

INTRODUCTION

1.1 THE EUROPEAN COMMUNITY

The formation of the European Community began in 1950 (NCBMP, 1988) when the French Minister of Foreign Affairs, Robert Schuman, proposed that European countries should pool their production and consumption of coal and steel and establish institutions to manage this. The first European community – the European Coal and Steel Community (ECSC) – was set up by a Treaty signed in Paris in April 1951. Two other European communities were established by the Treaties of Rome signed in March 1957. These were the European Economic Community (EEC) and the European Atomic Energy Community (Euratom). In 1986 these treaties were amended by the Single European Act, which was designed to improve the future working of the communities and to extend their scope.

The first Member States were Belgium, France, West Germany, Italy, Luxembourg and the Netherlands (the Six). Their Parliaments ratified the Treaty of Paris in 1951/52 and the Treaties of Rome in 1957. Denmark, Ireland and the UK became members in 1973. Greece entered the communities in 1981 and Portugal and Spain in 1986. Representatives of each of the Twelve signed the Single European Act in 1986.

Each of the Treaties established that the tasks entrusted to the ECSC, EEC and Euratom should be carried out by four institutions – a European Parliament, a Council, a Commission and a Court of Justice. Originally, the three communities had separate councils and commissions, but since 1967 there has been a single Council and a single Commission, which exercise the powers and responsibilities vested in their predecessors by the Treaties. The Parliament and the Court of Justice have always

been common to all three communities. As the three communities are managed by common institutions, they are generally referred to in the singular as the European Community (EC). The roles of the four institutions are briefly as follows.

The Commission ensures that the EC rules and principles are respected and proposes to the Council measures likely to advance the development of EC policies. The Council makes the major policy decisions of the EC and it can deal only with proposals from the Commission. The European Parliament does not have legislative powers – the Commission has the sole power of initiative and the Council plays the major role in taking decisions. The Parliament has an important role in three areas – adoption and control of the EC budget, consideration of proposals for EC legislation and general supervision over the activities of the institutions. The Court of Justice has the power to quash measures that are incompatible with the Treaties and can pass judgement on the interpretation or validity of points of EC law.

Since 1950, there has been continual progress with the idea of creating a common market within the EC, and in 1985, the Commission was asked by the Member States to put forward concrete proposals to achieve completion of a fully unified internal market by 1992. The Commission published its proposals in the form of a White Paper, which included a new approach to technical harmonization and standards.

1.2 TECHNICAL HARMONIZATION AND STANDARDS

For many years the EC attempted to remove technical barriers through the adjustment of national regulations to conform to an agreed EC standard.

This proved difficult and time-consuming and thus a new approach to technical harmonization was established in May 1985. The new approach adopts Community-wide standards for health and safety, which afford all Europeans with an equally high level of protection and leaves manufacturers whose products meet such standards the freedom to use their own manufacturing and design traditions and skills. It requires clear differentiation between these areas where harmonization is necessary and those which can be left to mutual recognition of national standards and regulations. EC policy on technical harmonization is established by a legal device known as a **Directive**. Of particular importance to the construction industry is the **Construction Products Directive** (CPD), which came into force on 27 December 1991 (DOE, 1991).

Its aim is to provide for the free movement, sale and use of construction products that are fit for their intended use and have such characteristics that structures in which they are incorporated meet certain **Essential Requirements**. The six Essential Requirements are:

1. Mechanical Resistance and Stability
2. Safety in Case of Fire
3. Hygiene, Health and Environment
4. Safety in Use
5. Protection Against Noise
6. Energy Economy and Heat Retention.

The broad statements of the essential requirements that are contained in the CPD are being expanded through a series of **Interpretative Documents** (ID). One such ID is concerned with the Essential Requirement 'Mechanical Resistance and Stability' (CEC, 1991), which is formulated in the CPD as follows:

> The construction works must be designed and built in such a way that the loadings are liable to act on it during its construction and use will not lead to any of the following:
> (a) collapse of the whole or part of the work
> (b) major deformations to an inadmissible degree
> (c) damage to other parts of the works or installed equipment as a result of major deformation of the load bearing construction
> (d) damage by an event to an extent disproportionate to the original cause.

It is further specified that:

> The products must be suitable for construction works which (as a whole and in their separate parts) are fit for their intended use, account being taken of economy, and in this connection satisfy the following essential requirements where the works are subject to regulations containing such requirements. Such requirements must, subject to normal maintenance, be satisfied for an economically reasonable working life. The requirements generally concern actions which are foreseeable.

The ID 'Mechanical Resistance and Stability' incorporates the **limit state concept** as a basis for verification and an essential part of the supporting documentation is a series of '**Structural Eurocodes**'.

1.3 STRUCTURAL EUROCODES

The Commission initiated the work of establishing a set of harmonized technical rules for the design of building and civil engineering works, which would initially serve as the alternative to the different rules in force in the various Member States and would ultimately replace them. These technical rules became known as the 'Structural Eurocodes' and work is in hand on the following, each generally consisting of a number of parts:

- EN 1991 Eurocode 1: Basis of design and actions on structures
- EN 1992 Eurocode 2: Design of concrete structures
- EN 1993 Eurocode 3: Design of steel structures
- EN 1994 Eurocode 4: Design of composite steel and concrete structures
- EN 1995 Eurocode 5: Design of timber structures
- EN 1996 Eurocode 6: Design of masonry structures
- EN 1997 Eurocode 7: Geotechnical design
- EN 1998 Eurocode 8: Design provisions for earthquake resistance of structures
- EN 1999 Eurocode 9: Design of aluminium structures.

In 1990, the Commission transferred work on further development, issues and updates of the Structural Eurocodes to the European Committee for Standardization (CEN). The CEN Technical Committee CEN/TC250 is responsible for all Structural Eurocodes. The Codes are intended to serve as reference documents for the following purposes:

1. As a means to prove compliance of building and civil engineering works with the essential requirements of the Construction Products Directive.

2. As a framework for drawing up harmonized technical specifications for construction products.

They cover execution and control only to the extent that it is necessary to indicate the quality of the construction products, and the standard of workmanship, needed to comply with the assumptions of the design rules. Until the necessary set of harmonized technical specifications for products and for methods of testing their performance is available, some of the Structural Eurocodes cover some of these aspects in annexes.

● Part 10: Fire resistance of concrete structures

4. Drafting not started
   ● Part 3: Concrete foundations and piling
   ● Part 4: Liquid-retaining structures
   ● Part 5: Temporary/short-design-life structures
   ● Part 6: Massive civil engineering structures.

ENV 206 Concrete – Performance, production, placing and compliance criteria, has been published as ENV 206 together with the UK national annex and progress is being made on standards for reinforcement and prestressing steel, cement, additions (used to define materials added at the concrete mixer, such as fly ash, silica fume and ground granulated blast furnace slag), aggregates, admixtures and mixing water.

There is no doubt that Eurocode 2 will meet fierce opposition from some quarters of the British construction industry, as did CP 110: 1972 and BS 8110: 1985. However, it must be remembered that general rules and rules for buildings in EC2 using the limit state concept originate in the pioneering work of the Comité Européen du Béton (CEB) dating back to the 1950s and, in particular, the 'International Recommendations for the Design and Construction of Concrete Structures', which was published in 1970, and was followed by the CEB Model Code in 1978. The limit state concept has now been fully established and forms the basis of the nine Eurocodes. The three-year validity of the ENV will allow adjustments to be made prior to conversion to a European Standard (EN)

EC2: Part 1 is broadly comparable with BS 8110: 1985 (Parts 1 and 2) except that, in EC2, precast concrete and lightweight concrete are covered in separate documents – Parts 1B and 1C respectively. There are some differences in terminology between BS 8110 and EC2, namely:

● Loads are referred to as **actions**
● Superimposed loads are **variable actions**
● Self-weight and dead loads are **permanent actions.**

1.5 LAYOUT OF EC2

The Code has seven chapters: (1) Introduction, (2) Basis of design, (3) Material properties, (4) Section and member design, (5) Detailing provisions, (6) Construction and workmanship and (7) Quality control. These are followed by four appendices covering time-dependent effects, non-linear analysis,

ures – Part
ldings, has
dard (ENV
er a period
√ period of
: supporting
Application
NAD is to
rticular in
be used for
NAD takes
sions in the

Approved
2-1-1: 1991
conjunction
lings. Com-
D does not
igations.
that all the
onstruction
European
e Eurocode
March 1993)

reinforced

ight aggre-

onded and
g tendons

ructures

prestressed

concrete bridges

additional design procedures for buckling and checking deflections by calculation.

A distinction is made between Principles and Application Rules. The **Principles** comprise general statements and definitions for which there is no alternative, together with requirements and analytical models for which no alternative is permitted unless specifically stated. The Principles are preceded by the letter P. The **Application Rules** are generally recognized rules that follow the Principles and satisfy their requirements. It is permissible to use alternative design rules different from the Application Rules given in the Code provided that it is shown that the alternative rules accord with the relevant Principles and are at least equivalent with regard to the resistance, serviceability and durability achieved for the structure with the present Code. Numerical values identified by being boxed are given as indications. Other values may be specified by Member States. It is assumed that the structures are designed by appropriately qualified and experienced personnel, that there is adequate supervision and quality control by personnel having the appropriate skill and experience, that the construction materials and products are in accordance with the relevant specifications, that the structure will be adequately maintained and that it will be used in accordance with the design brief.

As the purpose of this book is to introduce built environment students and graduates to the application of EC2 to the design of conventional reinforced concrete buildings, reference to the clauses on prestressed concrete has been omitted. In order to present the material in a format more suited to direct application to design, the sequence in which it is presented has, in part, been modified from that in EC2. Wherever practicable, use is made of simplified procedures, design charts and tables and the layout of calculations is under three main headings: Loading (actions), Member analysis and Section analysis.

1.6 FUNDAMENTAL REQUIREMENTS (CL. 2.1)

The fundamental requirements of EC2 related to the basis of design are given in full below:

P(1) A structure shall be designed and constructed in such a way that
- with acceptable probability, it will remain fit for the use for which it is required, having due regard to its intended life and its cost, and
- with appropriate degrees of reliability, it will sustain all actions and influences likely to occur during execution and use and

have adequate durability in relation to maintenance costs.

P(2) A structure shall also be designed in such a way that it will not be damaged by events like explosions, impact or consequences of human errors, to an extent disproportionate to the original cause.

P(3) The potential damage should be limited or avoided by appropriate choice of one or more of the following:
- avoiding, eliminating or reducing the hazards which the structure is to sustain
- selecting a structural form which has low sensitivity to the hazards considered
- selecting a structural form and design that can survive adequately the accidental removal of an individual element
- tying the structure together.

P(4) The above requirements shall be met by the choice of suitable materials, by appropriate design and detailing and by specifying control procedures for production, design, construction and use as relevant to the particular project.

1.7 LIMIT STATES (CL. 2.2.1.1)

Limit states are defined as the states beyond which the structure no longer satisfies the design performance. Limit states are classified into:

- Ultimate limit states
- Serviceability limit states.

Broadly, **ultimate limit states** (ULS) are associated with collapse, loss of equilibrium of the structure considered as a rigid body and failure by excessive deformation, rupture or loss of stability; and **serviceability limit states** (SLS) correspond to states beyond which specified service requirements are no longer met and include consideration of deformation or deflection, vibration, cracking of the concrete and the presence of excessive compressive stress.

A departure from BS 8110 is that EC2 requires a check on concrete compressive stress at service load. This check is to prevent formation of longitudinal cracks and microcracking in members and is covered in Chapter 4.

1.8 ACTIONS (CL. 2.2.2)

Actions are taken as:

- **Direct** actions, that is, a force (load) applied to a structure

● **Indirect** actions, that is, an imposed deformation such as temperature effects or settlement.

Actions are classified by their variation in time and their spatial variation. For the purposes of this introductory text, the following actions will be covered:

● **Permanent** actions (G), e.g. self-weight of structures, fittings, ancillaries and fixed equipment
● **Variable** actions (Q), e.g. imposed loads, wind loads or snow loads.

Seismic actions are covered in Chapter 7.

1.9 CHARACTERISTIC VALUES OF ACTIONS (CL. 2.2.2.2)

In EC1 (draft) (CEC, 1992), areas in buildings are divided into categories according to their specific use as shown in Table 1.1.

The corresponding characteristic values of the actions uniformly distributed (kN/m^2) and concentrated (kN) are given in Table 1.2 together with combination (ψ) factors. The combination factors ψ_0, ψ_1 and ψ_2 relate to rare, frequent and quasi-permanent load combinations respectively (see serviceability limit states, Chapter 4). The combination factors for the NAD are also listed in Table 1.3 and

it should be noted that the loading codes for the use of EC2 with the UK NAD are:

● BS 648: 1964   Schedule of weights of building materials
● BS 6399   Loading for buildings
● BS 6399: Part 1: 1984   Code of practice for dead and imposed loads
● BS 6399: Part 3: 1988   Code of practice for imposed roof loads
● CP 3   Code of basic data for the design of buildings
● CP 3: Chapter V   Loading
● CP 3: Chapter V: Part 2: 1972   Wind loads.

In using the above documents with EC2 the following modifications should be noted.

1. The imposed floor loads of a building should be treated as one variable action to which the reduction factors given in BS 6399: Part 1: 1984 are applicable.
2. Snow drift loads obtained from BS 6399: Part 3: 1988 should not be treated as accidental actions as defined in EC2. They should be multiplied by 0.7 and treated as a variable action.
3. The wind loading should be taken as 90% of the value obtained from CP 3: Chapter V: Part 2: 1972.

Table 1.1   EC1 (draft) – areas of buildings divided into categories

Category and uses	Examples
Category A Areas for domestic and residential activities	Rooms in residential buildings and houses Rooms and wards in hospitals Bedrooms in hotels and hostels Kitchens and toilets
Category B Areas where people may congregate 　(with the exception of areas defined under 　categories A, C and D)	Areas in public and administration buildings Offices Cafés, shops Areas in schools, barracks, hospitals, reformatories, 　prisons Areas in hotels, leisure centres, clubs
Category C Areas susceptible to overcrowding, including access 　areas	Assembly halls, churches, theatres, cinemas, concert 　halls, dance halls, gymnastic halls Conference rooms, lecture halls, exhibition rooms Restaurants, dining halls Reception halls, waiting rooms Platforms, stands, stages Shopping areas
Category D Areas susceptible to accumulation of goods, including 　access areas	Areas in warehouses Areas in department stores Areas in stationery and office stores

Table 1.2 Characteristic values of imposed loads on floors in buildings and ψ values. The local concentrated load shall be considered to act at any point of the floor or stairs and to have an application area comprising a square with a 50 mm side. Where the imposed loads from several storeys are relevant, the loads may be reduced by a reduction factor. Although the above loadings are broadly similar to those given in BS 6399, the relevant loadings for UK application should be taken from BS 6399. The NAD combination factors are given in Table 1.3

Loaded areas[a]	UDL (kN/m²)	Conc. load (kN)	ψ_0	ψ_1	ψ_2
Category A					
general	2.0	2.0	0.7	0.5	0.3
stairs	3.0	2.0	0.7	0.5	0.3
balconies	4.0	2.0	0.7	0.5	0.3
Category B					
general	3.0	2.0	0.7	0.5	0.3
stairs, balconies	4.0	2.0	0.7	0.5	0.3
Category C					
with fixed seats	4.0	4.0	0.7	0.7	0.6
other	5.0	4.0	0.7	0.7	0.6
Category D					
general	5.0	7.0	1.0	0.9	0.8

[a]See Table 1.1 for uses in each category.

Table 1.3 Combination factors for the NAD (table 1 of NAD)

Variable actions[a]	ψ_0	ψ_1	ψ_2
Imposed loads			
dwellings	0.5	0.4	0.2
offices and stores	0.7	0.6	0.3
parking	0.7	0.7	0.6
Wind loads	0.7	0.2	0
Snow loads	0.7	0.2	0

[a]For the purposes of EC2 these three categories of variable actions should be treated as separate and independent actions.

1.10 PARTIAL SAFETY FACTORS FOR ULTIMATE LIMIT STATE

A distillation of the partial safety factors listed in EC2 is given in Table 1.4. These are for the following actions on building structures:

1. **Persistent** situations corresponding to normal conditions of use of the structure
2. **Transient** situations, for example, during construction and repair.

In the NAD, it is assumed that the favourable effect corresponds to the lower value of the characteristic partial safety factor ($\gamma_{G, \text{inf.}} = 1.0$) and the unfavourable effect corresponds to the upper value of the characteristic partial safety factor ($\gamma_{G, \text{sup.}} = 1.35$). Thus Table 1.4 complies with the NAD.

Broadly, for the ultimate limit state, the use of Table 1.4 and the loading codes listed previously will comply with the requirements of EC2 and the NAD for application to building structures in the UK. The corresponding partial safety factors for materials (ULS) are:

$\gamma_c = 1.5$ concrete

$\gamma_s = 1.15$ steel reinforcement

1.11 SERVICEABILITY LIMIT STATES

EC2 refers to three combinations of actions for the serviceability limit state – rare combination, frequent combination and quasi-permanent combination. The Code defines these algebraically and a descriptive interpretation is given in Table 1.5. This is taken from notes to the revision of the draft of EC1: Part 2.4: Imposed loads.

In EC2, serviceability requirements for limiting compressive stress, deflection and crack width are generally based on the use of the quasi-permanent load combination, which is expressed as:

$$\sum G_{kj} + \sum \psi_{2,i} Q_{k,i} \text{ where } i \geq 1$$

where $G_{k,j}$ = characteristic value of permanent actions, $Q_{k,i}$ = characteristic value of variable actions, and ψ_2 = combination factor (see Table 1.2).

The use of quasi-permanent loading in serviceability checks is developed in Chapter 4.

Table 1.4 Partial safety factors for actions in building structures for persistent and transient design situations

Load combination	Permanent (γ_G)		Variable (γ_Q)		Wind
	Favourable effect	Unfavourable effect	Favourable effect	Unfavourable effect	
Permanent + variable	1.0	1.35	–	1.5	–
Permanent + wind	1.0	1.35	–	–	1.5
Permanent + variable + wind	1.0	1.35	–	1.35	1.35

Table 1.5 Descriptive interpretation of combination values

Representative value	Load	Load duration	Class	Example	Accumulated duration
Characteristic	Q_k	short-term	permanent	self-weight	more than 10 years
Rare combination	$\psi_0 Q_k$	medium-term	long-term	imposed	6 months to 10 years
Frequent combination	$\psi_1 Q_k$	long-term	medium-term	snow	1 week to six months
Quasi-permanent combination	$\psi_2 Q_k$	permanent	short-term	wind	less than 1 week
–	–	–	instantaneous	accidental	–

1.12 MATERIAL PROPERTIES

1.12.1 Normal weight concrete (cl. 3.2.1)

Normal weight concrete is defined as having an oven-dry (105°C) density greater than 2000 kg/m³, but not exceeding 2800 kg/m³. Properties of normal weight concrete for use in design are based on the 28-day characteristic compressive cylinder strength (f_{ck}). Note that the classification of concrete (for example, C25/30) refers to cylinder/cube strength. In the absence of more accurate data, the properties of concrete can be derived from the following equations.

The mean value of the tensile strength f_{ctm}, noting that the term 'tensile strength' relates to the maximum stress that the concrete can withstand when subjected to uniaxial tension, is given by:

$$f_{ctm} = 0.3 f_{ck}^{2/3} \qquad \text{(EC2 eqn. 3.2)}$$

$$f_{ctk,0.05} = 0.7 f_{ctm} = 0.21 f_{ck}^{2/3} \qquad \text{(EC2 eqn. 3.3)}$$

$$f_{ctk,0.95} = 1.3 f_{ctm} = 0.39 f_{ck}^{2/3} \qquad \text{(EC2 eqn. 3.4)}$$

where $f_{ctk,0.05}$ is the lower characteristic tensile strength (5% fractile) and $f_{ctk,0.95}$ is the upper characteristic tensile strength (95% fractile). The basic design shear strength of concrete (τ_{Rd}) (see Chapter 4) is defined by:

$$\tau_{Rd} = 0.25 f_{ctk,0.05}/\gamma_c \qquad \text{(EC2 clause 4.3.2.1)}$$

where $\gamma_c = 1.5$. Thus

$$\tau_{Rd} = 0.035 f_{ck}^{2/3}$$

The secant modulus of elasticity of normal weight concrete is defined by:

$$E_{cm} = 9.5(f_{ck} + 8)^{1/3} \text{ kN/mm}^2 \qquad \text{(EC2 eqn. 3.5)}$$

In the above equations, f_{ck} is expressed in N/mm². The values obtained for E_{cm} relate to concrete cured under normal conditions and made with aggregates predominantly consisting of quartzite (metamorphosed sandstone) gravel. Design values for the ultimate bond stress f_{bd} in conditions of good bond (see Chapter 4) are given by:

$$f_{bd} = [0.36(f_{ck})^{1/2}]/\gamma_c \quad \text{plain bars} \qquad \text{(EC2 eqn. 5.1)}$$

$$= (2.25 f_{ctk,0.05})/\gamma_c \quad \text{high-bond bars} \qquad \text{(EC2 eqn. 5.2)}$$

For convenience, the properties of concrete referred to above have been related to the nine strength classes and are given in Table 1.6. Further properties of concrete are:

● **Stress – strain diagram** for concrete in uniaxial compression (see Chapter 4).

Table 1.6 Summary of properties of concrete all related to the characteristic compressive cylinder strength of concrete (f_{ck}, N/mm^2) at 28 days

Strength class of concrete	C12/15	C16/20	C20/25	C25/30	C30/37	C35/45	C40/50	C45/55	C50/60	Formulae
f_{ck}	12	16	20	25	30	35	40	45	50	–
f_{ctm}	1.6	1.9	2.2	2.6	2.9	3.2	3.5	3.8	4.1	$0.3f_{ck}^{2/3}$ (N/mm^2)
$f_{ctk,0.05}$	1.1	1.3	1.5	1.8	2.0	2.2	2.5	2.7	2.9	$0.21f_{ck}^{2/3}$(N/mm^2)
$f_{ctk,0.95}$	2.0	2.5	2.9	3.3	3.8	4.2	4.6	4.9	5.3	$0.39f_{ck}^{2/3}$(N/mm^2)
τ_{Rd}	0.18	0.22	0.26	0.30	0.34	0.37	0.41	0.44	0.48	$0.035f_{ck}^{2/3}$ (N/mm^2)
E_{cm}	26	27.5	29	30.5	32	33.5	35	36	37	$9.5\,(f_{ck}+8)^{2/3}$ (kN/mm^2)
f_{bd} plain bars	0.9	1.0	1.1	1.2	1.3	1.4	1.5	1.6	1.7	$[0.36(f_{ck})^{1/2}]/\gamma_c$
f_{bd} high-bond bars $\varphi \geq 34$ mm	1.6	2.0	2.3	2.7	3.0	3.4	3.7	4.0	4.3	$(2.25f_{ctk,0.05}^{\wedge}>)/\gamma_c$

Table 1.7 Differences between current British Standards and prEN 10080 (table 5 of NAD)

Property	BS 4449 and BS 4483	prEN 10080
Specified characteristic yield strength	grade 460 N/mm^2 grade 250 N/mm^2	500 N/mm^2 not included
Bond strength for:		
ribbed bars/wires	deformed type 2	high bond
indented wires	deformed type 1	not included
plain bars/wires	plain rounds	not included
Ductility class[a] (now defined as elongation at maximum load and ultimate to yield strength ratio)	not covered	class H or class N (this may be deleted in the final version)[b]

[a]In design where plastic analysis or moment distribution over 15% is used, it is essential to specify ductility class H as defined in prEN 10080 since this parameter is not covered by BS 4449 and BS 4483.

[b]All ribbed bars and all grade 250 bars may be assumed to be class H. Ribbed wire welded fabric may be assumed to be available in class H in wire sizes of 6 mm or over. Plain or indented wire welded fabric may be assumed to be available in class N.

- **Poisson's ratio** – for design purposes, Poisson's ratio for elastic strains may be taken as 0.2 (cl. 3.1.2.5.3 (P1)); if cracking is permitted for concrete in tension, it may be assumed as zero (P2).
- **Coefficient of thermal expansion** – for design purposes, where thermal expansion is not of great influence, it may be taken as equal to $10 \times 10^{-6}/°C$ (cl. 3.1.2.5.4 (P2)).

1.12.2 Reinforcing steel (cl. 3.2)

Section 3.2 (reinforcing steel) of EC2 gives properties of reinforcement for use in structural concrete for which Euronorm (EN 10080 for reinforcement) is currently being drafted. The differences between current British Standards and prEN 10080 are given in table 5 of the NAD, which is reproduced here as Table 1.7.

EC2 defines two ductility classes and it can be seen from Table 1.7 that this is not covered by the British Codes. The two ductility classes are:

- **high ductility** (H) for which

$$e_{uk} > \boxed{5}\%: \quad \text{value of } (f_t/f_y)_k > \boxed{1.08}$$

- **normal ductility** (N) for which

$$e_{uk} > \boxed{2.5}\%: \quad \text{value of } (f_t/f_y)_k > \boxed{1.05}.$$

Here f_{tk} is the characteristic tensile strength of reinforcement, f_{yk} is the characteristic yield strength of the reinforcement, and e_{uk} is the characteristic elongation of the reinforcement at maximum.

The matter is further complicated by the draft of EN 1998 Eurocode 8: Design provisions for earthquake resistance of structures, which calls for three ductility classes depending on whether the building is being designed for low, medium or high ductility (see Chapters 4 to 7). It is envisaged that eventually there will be three grades of reinforcement: normal ductility, high ductility and a seismic grade. Further, it is likely that when EN 10080 is published, grade 500 will replace grade 460. Reference should be made to the note at the bottom of Table 1.7 regarding design where plastic analysis or moment redistribution over 15% is used. It will normally be the case, but it should be checked with reinforcement manufacturers, that ductility class H is complied with. Care should be exercised with serviceability checks when using steel grades higher than S400 as basic span/effective depth ratios to limit deflections given in EC2 and the NAD correspond to f_{yk} of about 400 N/mm² (see Chapter 4).

Other properties of reinforcement are:

- Density: 7850 kg/m³.
- Modulus of elasticity: 200 kN/mm² (mean).
- Coefficient of thermal expansion: $10 \times 10^{-6}/°C$.

1.13 SUMMARY

For the application of EC2 in conjunction with the UK NAD to the design of conventional reinforced concrete buildings, the following procedure is recommended:

1. Establish characteristic loadings from BS 648, BS 6399, CP 3 – refer to NAD.
2. Select partial safety factors for permanent and variable actions (ultimate limit state) from Table 1.4.

3. Establish durability requirements from Chapter 2.
4. Adopt concrete grade from the range C25/30, C30/37 and C35/45 depending on durability requirements.
5. Adopt steel grade 460 to BS 4449/BS 4483. Where plastic analysis or moment redistribution over 15% is used, specify ductility class H.
6. For loading arrangements and member analysis, see Chapter 3.
7. For section analysis (ULS and SLS), see Chapters 4 and 5.

REFERENCES

CEC (1991) *Interpretative Document for the Essential Requirement, Mechanical Resistance and Stability*, Commission of the European Communities, Technical Committee 89/106/TC1, Document TC1/018-Rev. 1, Brussels, July.

CEC (1992) *Eurocode 1: Basis of Design and Actions on Structures*, CEN/TC250/SC1/1992 – Draft, September.

DOE (1991) *European Construction*, Issue No. 15, Special Supplement – *Construction Products Directive*, DOE Construction Policy Directive and *Building Magazine*, September. (For further information, contact the Secretariat, Construction Directive, Room P1/111, 2 Marsham Street, London SW1P 3EB.)

NCBMP (1988) *The UK Construction Industry and the European Community*, National Council of Building Materials Producers and *Building Magazine*, June.

2

DESIGN
FOR DURABILITY

2.1 INTRODUCTION

Lack of durability of reinforced concrete structures is a worldwide problem involving an annual expenditure of millions of pounds on inspections, maintenance and repairs. This can be largely attributed to inadequate attention to durability at the design and construction stage of a project. Design and construction should be properly integrated at the initial stages of a building project, with emphasis not only on strength, stability, cost and buildability, but also on durability. The proportioning of structural members for minimum structural depth to meet strength requirements with consequent congestion of reinforcement will generally have an adverse effect on buildability, leading to a greater risk of defects and deterioration becoming apparent within a few years after completion of construction. Recent Codes of Practice, e.g. CP 110 (1972), BS 8110 (1985) and EC2 (1992), have paid increasing attention to design for durability, and EC2 (cl. 2.4) states that to ensure an adequately durable structure, the following interrelated factors shall be considered:

- The expected environmental conditions
- The use of the structure
- The required performance criteria
- The composition, properties and performance of the materials
- The shape of members and structural detailing
- The quality of workmanship and level of control
- The particular protective measures
- The likely maintenance during the intended life.

2.2 ENVIRONMENTAL CONDITIONS

A requirement of EC2 is that environmental conditions shall be estimated at the design stage to assess their significance in relation to durability and to enable adequate provision to be made for protection of the materials. Broadly, the climate of Europe can be classified as **humid mesothermal** (Trevatha, 1961), that is:

1. Dry summer subtropical (central and southern Spain, southern France, central and southern Italy and Greece)
2. Humid subtropical, warm summer (northern Italy)
3. Marine, cool summer (northern Spain, central and northern France, Germany, Belgium, the Netherlands and the United Kingdom).

Table 2.1 gives a general picture of design temperatures and precipitation in 11 European Community capital cities (IHVE, undated). Other considerations are wind and atmospheric pollution, grit and dust, smoke gases and motor vehicle exhaust. Environmental considerations will influence building orientation, structural configuration, surface treatments and durability. As with BS 8110, EC2 relates a series of exposure classes to environmental conditions. These are reproduced in Tables 2.2 and 2.3. It is important to consider how mechanisms of deterioration of reinforced concrete are influenced by environmental factors.

Table 2.1 Design temperatures and precipitation in 11 European Community capital cities

Country	Lat.	Long.	Height (m)	Month	Dry bulb	Wet bulb	Average diurnal range (°C)	Max.	Min.	Annual	Max. in 24 h.
					Design temp. (°C)			Precipitation (mm)			
								Average monthly			
Belgium (Brussels)	50° 48' N	04° 21' E	100	July	31	21	11	89	51	838	48
Denmark (Copenhagen)	55° 41'N	12° 33' E	13	July	28	20	9	81	30	592	41
France (Paris)	48° 49' N	02° 29' E	50	July	32	21	12	51	33	566	51
Germany (Berlin)	52° 27' N	13° 18' E	57	July	32	21	11	79	33	587	66
Irish Republic (Dublin)	53° 22' N	06° 21' W	47	July	25	17	9	76	48	754	69
Netherlands (Amsterdam)	52° 23' N	04° 55' E	2	July	28	19	6	71	33	680	58
Portugal (Lisbon)	38° 31' N	09° 08' W	95	August	34	22	9	107	5	686	112
Greece (Athens)	37° 58' N	23° 43' E	107	July	37	22	10	71	5	401	150
Italy (Rome)	41° 48' N	12° 36' E	115	August	36	23	12	97	15	653	–
Spain (Madrid)	40° 25' N	03° 41' W	666	July	36	22	14	56	8	419	66
United Kingdom (London)	51° 28' N	00° 19' W	5	July	28	19	9	64	36	580	56

Table 2.2 Environment and exposure condition classes (table 3.2 of BS 8110: 1985)

Environment	Exposure conditions
Mild	Concrete surfaces protected against weather or aggressive conditions
Moderate	Concrete surfaces sheltered from severe rain or freezing whilst wet
	Concrete subject to condensation
	Concrete surfaces continuously under water
	Concrete in contact with non-aggressive soil (table 6.1 – class 1)[a]
Severe	Concrete surfaces exposed to severe rain, alternate wetting and drying or occasional freezing or severe condensation
Very severe	Concrete surfaces exposed to sea water spray, de-icing salts (directly or indirectly), corrosive fumes or severe freezing conditions whilst wet
Extreme	Concrete surfaces exposed to abrasive action, e.g. sea water carrying solids or flowing water with pH ≤ 4.5 or machinery or vehicles

[a]For aggressive soil conditions see BS 8110 (cl. 6.2.3.3).

2.3 ENVIRONMENTAL FACTORS

Critical environmental factors are: the concentration of **carbon dioxide** (CO_2) in the air; the presence of **chloride ions** (from de-icing salts or sea water); the presence of **moisture** (H_2O) and **oxygen** (O_2); and the **temperature**. In order to appreciate the influence of environmental factors on the durability of concrete, the designer should be made aware of the constituents of cement and the reaction that takes place between cement and water (hydration). Table 2.4 lists the constituents of ordinary Portland cement (OPC) and summarizes the basic reactions with water and the atmosphere. The major products of the reaction of cement with water are calcium silicate hydroxide (abbreviated $C_3S_2H_3$) and calcium hydroxide $Ca(OH)_2$. The hydration product calcium hydroxide forms a protective alkaline environment to the reinforcement, and to prevent corrosion the alkaline environment must be maintained.

2.4 THE C$_3$S/C$_2$S RATIO

In modern cements, the **C$_3$S/C$_2$S ratio** has increased and they are more finely ground. This has resulted

Table 2.3 Environmental conditions and exposure classes (table 4.1 of EC2 (ENV 1992-1-1), table 6.2.1 in ENV 206)

Exposure class		Examples of environmental conditions
1 Dry environment		Interior of buildings for normal habitation or offices[a]
2 Humid environment	**a** Without frost	Interior of buildings where humidity is high (e.g. laundries)
		Exterior components
		Components in non-aggressive soil and/or water
	b With frost	Exterior components exposed to frost
		Components in non-aggressive soil and/or water and exposed to frost
		Interior components when the humidity is high and exposed to frost
3 Humid environment with frost and de-icing salts		Interior and exterior components exposed to frost and de-icing agents
4 Sea water environment	**a** Without frost	Components completely or partially submerged in sea water, or in the splash zone
		Components in saturated salt air (coastal area)
	b With frost	Components partially submerged in sea water or in the splash zone and exposed to frost
		Components in saturated salt air and exposed to frost

The following classes may occur alone or in combination with the above classes:

5 Aggressive chemical environment[b]	**a** Slightly aggressive chemical environment (gas, liquid or solid)
	Aggressive industrial atmosphere
	b Moderately aggressive chemical environment (gas, liquid or solid)
	c Highly aggressive chemical environment (gas, liquid or solid)

[a]This exposure class is valid only as long as during construction the structure or some of its components is not exposed to more severe conditions over a prolonged period of time.
[b]Chemically aggressive environments are classified in ISO/DP 9690. The following equivalent exposure conditions may be assumed:
Exposure class 5a ISO classification A1G, A1L, A1S.
Exposure class 5b ISO classification A2G, A2L, A2S.
Exposure class 5c ISO classification A3G, A3L, A3S.

Table 2.4 Summary of the constituents of ordinary Portland cements and basic reactions with water and the atmosphere (CO_2)

Name of compound	Oxide composition	Abbreviation	Cement range typical for UK (wt%)	Hydration – reaction with water (exothermic)	Carbonation – CO_2 in atmosphere
Tricalcium silicate	$3CaO.SiO_2$	C_3S	42–67	$2C_3S+6H_2O=C_3S_2H_3 + 3Ca(OH)_2$ $Ca(OH)_2$ provides alkaline environment	$Ca(OH)_2 + CO_2 = CaCO_3 + H_2O$ Alkaline environment destroyed
Dicalcium silicate	$2CaO.SiO_2$	C_2S	8–31	$2C_2S + 4H_2O = C_3S_2H_3 + Ca(OH)_2$ $Ca(OH)_2$ provides alkaline environment	$Ca(OH)_2 + CO_2 = CaCO_3 + H_2O$ Alkaline environment destroyed
Tricalcium aluminate	$3CaO.Al_2O_3$	C_3A	5–14	Attacked by salts (e.g. sulphates in soil): sulphate-resisting cement characterized by a low percentage of C_3A	
Tetracalcium aluminoferrite	$4CaO.Al_2O_3.Fe_2O_3$	C_4AF	6–12	Comparatively inactive	
Free lime		C	1.7		
Alkalis			0.61	Alkali aggregate reaction – requires sufficient alkali, moisture and silica in the aggregate – risk above 0.6 % alkali	
Others			5.0		

in a marked increase in the strength of the concrete, and this strength increase is most noticeable at an early age. Whilst longer-term strength has also increased, the proportion of this strength achieved after 28 days has probably decreased. Thus it appears that the hydration process stops much earlier as the rate of hydration is much higher. The increase in strength is also accompanied by an increase in the early heat of hydration. If concrete is designed to a strength specification only, then specified strengths can now be met with much lower cement contents and increased water/cement ratios. If the increase in cement strength is used to reduce cement content and increase the water/cement ratio, the permeability and hence the durability of present-day concrete is likely to be poorer. This problem is firmly addressed in BS 8110 and EC2/NAD as concrete grades are related to the cement content and water/cement ratio. However, there is still a problem with regard to chloride penetration and this is discussed in section 2.7.

2.5 CARBONATION

Concrete is a porous material and the carbon dioxide CO_2 in the atmosphere may therefore penetrate via the pores into the interior of the concrete. A chemical reaction will take place with the calcium hydroxide $Ca(OH)_2$, which in simplified terms can be expressed as:

$$Ca(OH)_2 + CO_2 = CaCO_3 + H_2O$$

It is mainly the $Ca(OH)_2$ that influences the alkalinity of the concrete, which can rise to a pH value greater than 12.5. At this pH level, a microscopic oxide layer is formed on the steel surface – a passive film – which impedes the dissolution of the iron. If the concrete **carbonates**, the alkaline environment is destroyed, and in the presence of moisture and oxygen, the reinforcement will inevitably corrode. The iron will be oxidized by the atmospheric oxygen and then react with the water to form the corrosion product $Fe_2O_3(H_2O)$ known as hydrated iron oxide (rust). The corrosion product occupies a much greater volume than the metal from which it was formed, and thus sets up bursting forces in the surrounding concrete, leading to cracking and spalling. Roughly simplified, the **rate of carbonation** follows a square root time law, which can be expressed in the form:

$$\text{depth of carbonation} = k(t)^{1/2} \qquad (2.1)$$

where k is a coefficient and t is time expressed in years. The coefficient k is dependent on a number

Table 2.5 Influence of w/c on carbonation depth for OPC (no additives) aggregate type, sand and gravel

w/c ratio	Carbonation time (years) for cover (mm)					
	5	10	15	20	25	30
0.45	19	75	100+	100+	100+	100+
0.50	6	25	56	99	100+	100+
0.55	3	12	27	49	76	100+
0.60	1.8	7	16	27	45	65
0.65	1.5	6	13	23	36	52
0.70	1.2	5	11	19	30	43

Table 2.6 Some examples of carbonation depth

Depth of carbonation (mm)	Age of concrete (years)	Environment	Location
5	80+	damp, inland exposed	near Dorking, Surrey (UK)
30	12	exposed, inland	Cambridge (UK)
30	20	exposed, marine	Near Athens (Greece)

of factors, but of major importance are the water/cement ratio (w/c), compaction and curing. A number of empirical formulae (Nishi, 1962; Browne, 1986; Wallbank, 1989; CEIB, 1992; Parrott, 1987) have been proposed for estimating carbonation depth.

Table 2.5, developed from equations in Nishi (1962), which should be considered as indicative only, demonstrates the importance of cover and water/cement ratio in relation to the time for the carbonation front to reach the level of the reinforcement.

Thus from Table 2.5, for a water/cement ratio of 0.55 and a substandard cover of, say, 10 mm, the time for the carbonation front to reach the level of the reinforcement will be 12 years. Carbonation depths can be extremely variable depending on environmental factors and the quality of the concrete cover. Some examples are given in Table 2.6.

2.6 COVER TO REINFORCEMENT

As with BS 8110: 1985, EC2 relates cover requirements to exposure class, cement content and water/cement ratio. For a comparison of cover requirements for normal weight concrete, refer to Tables 2.7 (table 3.4 of BS 8110), 2.8 (table 6 of NAD and table 4.2 of EC2) and 2.9 (table 3 of ENV

Table 2.7 Nominal cover to all reinforcement (including links) to meet durability requirements for various conditions of exposure, water/cement ratio, cement content and concrete grade (table 3.4 of BS 8110)[a]

Conditions of exposure (see cl. 3.3.4)	Nominal cover (mm)				
Mild	25	20	20[b]	20[b]	20[b]
Moderate	–	35	30	25	20
Severe	–	–	40	30	25
Very severe	–	–	50[c]	40[c]	30
Extreme	–	–	–	60[c]	50
Maximum free water/cement ratio	0.65	0.60	0.55	0.50	0.45
Minimum cement content	275	300	325	350	300
Lowest grade of concrete (kg/m³)	C30	C35	C40	C45	C50

[a]This table relates to normal weight aggregate of 20 mm nominal maximum size. For concrete used in foundations to low-rise construction, see cl. 6.2.4.1.
[b]These covers may be reduced to 15 mm provided that the nominal maximum size of aggregate does not exceed 15 mm.
[c]Where concrete is subjected to freezing whilst wet, air entrainment should be used (see cl. 3.3.4.2).

Table 2.8 Minimum cover requirements for normal weight concrete with reinforcement, from NAD and EC2

(a) NAD (DD ENV 1992-1-1: 1992, table 6)[a]

		Exposure class, according to table 4.1								
		1	2a	2b	3	4a	4b	5a	5b	5c
Reinforcement	Nominal[b]	20	35	35	40	40	40	35	35	45
		(15)	(30)	(30)	(35)	(35)	(35)	(30)	(30)	(40)

[a]In order to satisfy the provisions of cl. 4.1.3.3 P(3) these values for cover should be associated with particular concrete qualities, to be determined from table 3 of ENV 206 and its National Annex. A reduction of 5 mm may be made where concrete of strength class C40/50 and above is used for reinforced concrete in exposure classes 2a to 5b. For slab elements, a further reduction of 5 mm may be made for exposure classes 2 to 5. For exposure class 5c a protected barrier should be provided to prevent direct contact with aggressive media.
[b]The nominal values for cover have been obtained from the minimum values allowing for a negative construction tolerance of 5 mm.

(b) EC2 (ENV 1992-1-1: 1991, table 4.2)[c]

		Exposure class, according to table 4.1								
		1	2a	2b	3	4a	4b	5a	5b	5c
Minimum cover (mm)	Reinforcement	15	20	25	40	40	40	25	30	40

[c]In order to satisfy the provisions of cl. 4.1.3.3 P(3) these minimum values for cover should be associated with particular concrete qualities, to be determined from table 3 in ENV 206. For slab elements, a reduction of 5 mm may be made for exposure classes 2–5. A reduction of 5 mm may be made where concrete of strength class C40/50 and above is used for reinforced concrete in exposure classes 2a–5b. However, the minimum cover should never be less than that for exposure class 1 in table 4.2. For exposure class 5c, the use of a protective barrier, to prevent direct contact with the aggressive media, should be provided.

206: 1990). In table 3 of ENV 206, reference is made to impermeable concrete (cl. 7.3.1.5) In clause 7.3.1.5, a mix is considered as suitable for water-impermeable concrete if the resistance to water penetration when tested according to ISO 7031 (Concrete hardened – Determination of the depth of penetration of water under pressure, as amended in Annex A of ENV 206) results in maximum values of penetration less than 50 mm and mean average values of penetration less than 20 mm. The water/cement ratio shall not exceed 0.55. Cover requirements to EC2/NAD and BS 8110 are compared in the following example. External columns are in an environment exposed to severe rain, alternate wetting and drying or occasional freezing or severe condensation. In accordance with table 3.2 of BS 8110, this would be classified as a severe environment and from table 3.4 a grade 40 concrete would be specified with a maximum free water/cement ratio of 0.55, a minimum cement content of 325 kg/m³ and a nominal cover of 40 mm. Referring to table 4.1 of EC2, the equivalent durability class could be taken as 2(b) and from the NAD (table 6) and ENV 206 (table 3) the maximum water/cement ratio is 0.55 and the minimum cement content is 280 kg/m³. Thus it would appear that, for this particular case, BS 8110 requirements for durability are more onerous than EC2/NAD.

Table 2.9 Durability requirements[a] related to environmental exposure for reinforced concrete (table 3 of ENV 206: 1990). See ENV 206 for plain and prestressed concrete

Requirements	Exposure class according to table 2 of ENV 206 or table 4.1 of EC2								
	1	2a	2b	3	4a	4b	5a	5b	5c[b]
Max w/c ratio for reinforced concrete[c]	0.65	0.60	0.55	0.50	0.55	0.50	0.55	0.50	0.45
Min. cement content (kg/m³) for reinforced concrete[c]	260	280	280	300	300	300	280	300	300
Min. air content of fresh concrete (%) for nominal max. aggregate size[d] of									
32 mm	–	–	4[e]	4[e]	–	4[e]	–	–	–
16 mm	–	–	5[e]	5[e]	–	5[e]	–	–	–
8 mm	–	–	6[e]	6[e]	–	6[e]	–	–	–
Frost-resistant aggregates[f]	–	–	yes	yes	–	yes	–	–	–
Impermeable concrete according to clause 7.3.1.5	–	–	yes	yes	yes	yes	yes	yes	yes
Types of cement for plain and reinforced concrete according to EN 197							sulphate-resisting cement[g] for sulphate contents > 500 mg/kg in water > 3000 mg/kg in soil		

[a]These values of w/c ratio and cement content are based on cement where there is long experience in many countries. However, at the time of drafting this prestandard, experience with some of the cements standardized in EN 197 is limited to local climatic conditions in some countries. Therefore during the life of this prestandard, particularly for exposure classes 2b, 3 and 4b, the choice of the type of cement and its composition should follow the national standards or regulations valid in the place of use of the concrete. Alternatively, suitability for the use of the cements may be proved by testing the concrete under the intended conditions of use.

[b]In addition, the concrete shall be protected against direct contact with the aggressive media by coatings unless for particular cases such protection is considered unnecessary.

[c]For minimum cement content and maximum water/cement ratio laid down in this standard, only cement listed in clause 4.1 (cements: Portland cement (CE1), Portland and composite cement (CE11), blast furnace cement (CE111) and pozzolanic cement (CE1V) shall comply with EN 197 Parts 1–3; other cements shall comply with the national standards or regulations valid in the place of use of the concrete) shall be taken into account. When pozzolanic or latent hydraulic additions are added to the mix, national standards or regulations, valid in the place of use of the concrete, may state if and how the minimum and maximum values respectively are allowed to be modified.

[d]With a spacing factor of air-entrained void system less than 0.2 mm measured on the hardened concrete.

[e]In cases where the degree of saturation is high for prolonged periods of time. Other values or measures may apply if the concrete is tested and documented to have adequate frost resistance according to the national standards or regulations valid in the place of use of the concrete.

[f]Assessed against the national standards or regulations valid in the place of use of the concrete.

[g]The sulphate resistance of the cement shall be judged on the basis of national standards or regulations valid in the place of use of the concrete.

Table 2.10 Potential for corrosion based on the *BRE Digest* No. 264

Chloride ion content as a percentage of cement content	Carbonation	
	Less than cover depth	Greater than cover depth
Low (up to 0.4)	low risk in all environmental conditions	moderate risk in damp conditions
Medium (0.4–1.0)	moderate risk in damp conditions	high risk enhanced by damp conditions and poor-quality concrete
High (above 1.0)	high risk enhanced by damp conditions and poor-quality concrete	high risk enhanced by damp conditions and poor-quality concrete

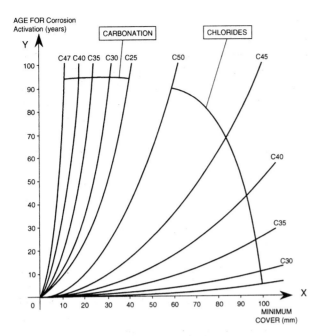

Figure 2.1 Prediction of time to corrosion activation (chlorides and carbonation). After Browne (1987).

2.7 CHLORIDES

The potential for corrosion of reinforcement is enhanced if chloride ions are present in the concrete. In BS 8110 and EC2/NAD it is stated that calcium chloride-based admixtures should not be added to reinforced concrete, prestressed concrete and concrete containing embedded metal. The limit of chloride ions (Cl⁻) by mass of cement for reinforced concrete in EC2/NAD and BS 8110 is 0.4% (0.2% for cement complying with BS 4027 and BS 4248). The potential for corrosion of reinforcement is summarized in Table 2.10, which *is* based on *BRE Digest* No. 264 (BRE, 1982). The presence of chloride ion concentration in the medium- to high-risk category will normally manifest itself within a few years, particularly if it is associated with concrete of low cement content and high permeability in a moist environment. A suggested basis for a design chart for durability, after Dr R.D. Browne, Taywood Engineering Ltd (Browne, 1987), which relates age for corrosion activity, minimum cover and concrete grade to carbonation and chlorides, is shown in Figure 2.1. The need to specify high-strength concrete with the appropriate constituents and adequate cover is immediately apparent. The curves on Figure 2.1 were calculated for UK conditions and, for chlorides in particular, it should be noted that increasing the strength grade from C30 to C50 shifts the curve significantly towards the *Y*-axis. Recent research (NCE, 1993) has indicated that concrete made with modern Portland cement is somewhat

Table 2.11 Figures relating free water/cement ratio to time required for capillaries to be blocked

Free w/c ratio	Time
0.4	3 days
0.45	7 days
0.5	14 days
0.6	6 months
0.7	1 year

inadequate in resisting chloride ion penetration unless a grade 60 or greater concrete is specified, together with cover to the reinforcement not less than 100 mm. For a salt-laden environment, ordinary Portland cement (OPC) blended with ground granulated blast furnace slag (GGBS), pulverized fly ash (PFA) or microsilica will give enhanced resistance to chloride ion penetration at a lower strength grade and cover. It appears that mixes in which the cementitious binder reacted more slowly and continued to react over a number of years resisted chloride ion penetration far better than those in which the hydration process was virtually exhausted in a few months. The matter is still under consideration by the European Standards Drafting Body (CEN).

2.8 CURING

As we have seen in section 2.4, hydration of Portland cements is a complex chemical reaction between cement minerals and water, the understanding of which is continuing to develop. The classic work of Powers *et al.* in the 1950s (e.g. Powers *et al.*, 1959) established that, for cement pastes, the time required for the hydration reactions to proceed to the level of blocking the capillaries and thus producing discontinuity and low porosity is dependent on the water/cement ratio.

Typical figures are given in Table 2.11 relating free water/cement ratio to time required for capillaries to be blocked. This table indicates the significance of w/c ratios in relation to curing time. However, these results were for cement pastes and a cement composition that differed from modern cements. The views of a Concrete Society study group on curing have recently been published (Cather, 1992) and it is stated that to use a parameter of capillary discontinuity is probably over-conservative. Further, from NCE (1993), it appears that, in an environment containing chlorides, the better the cure, the higher the chloride level near the surface of the concrete. Chlorides penetrate the cover zone by means of an absorption process for the first 15 mm or so. Further penetration is by a diffusion process, and it is argued that, with ordinary Portland cements, the reduction in effective diffusion coefficient that can be achieved by good curing does not appear to be sufficient to offset the increased chloride levels in the outer 15 mm of concrete. However, for blended cements, the diffusion coefficient can be an order of magnitude lower. Prior to the specification of a curing regime, the contents of Cather (1992) and NCE (1993) should be considered, but further research and discussion will be necessary before any modification of the BS 8110 and EC2/NAD recommendations can be made. These recommendations broadly require a *minimum*, but varying, time period during which curing procedures should be maintained. In BS 8110, table 6.5, the minimum periods of curing and protection for an average surface temperature of concrete between 5 and 25°C are related to the type of cement, the ambient conditions after casting (poor, average and good) and the average surface temperature of the concrete. In good ambient conditions after casting (relative humidity greater than 80%, protected from sun and wind) there are no special requirements for all cements – this has been liberally interpreted on construction sites even though clauses 6.6.1/2 of BS 8110 express clearly the importance of curing. Again, ENV 206 stresses the importance of thorough curing and protection for an adequate period. Clause 10.6.3 states that the required curing time depends on the rate at which a certain impermeability (resistance to penetration of gases or liquids) of the surface zone (cover to the reinforcement) of the concrete is reached. Curing times shall be determined by one of the following:

- From the maturity based on degree of hydration of the concrete mix and ambient conditions
- In accordance with local requirements
- In accordance with the minimum periods given in table 12 (ENV 206).

In table 12 of ENV 206, minimum curing times in days for exposure classes 2 to 5a depend on the rate of strength development of concrete (rapid, medium and slow, governed by *w/c* ratio; see table 13 of ENV 206), temperature of concrete during curing and ambient conditions during curing.

2.9 CEMENT CONTENT

In BS 8110 and ENV 206, minimum cement contents (kg/m³) are related to the conditions of exposure, nominal cover, water/cement ratio and concrete grade. The risk to durability of reducing the cement content and increasing the water/cement ratio has been emphasized previously, but it is also important

to consider the significance of cement content in relation to alkali–silica reaction (ASR). This is covered in some detail in clause 6.2.5.4 of BS 8 110. Some aggregates containing particular varieties of silica may be susceptible to attack by alkalis (Na_2O and K_2O) originating from cement (see Table 2.4) and other sources, producing an expansive action that can damage the concrete. This reaction will normally occur when *all* of the following are present:

1. High moisture level within the concrete
2. Cement with a high alkali content or other source of alkali
3. Aggregate containing alkali reactive constituents.

It is important to establish, as far as possible, the service records of the cement/aggregate combinations proposed for the project and to check that there have not been instances of alkali–silica reaction. If the materials are unfamiliar, precautions can take the following form:

1. Measures to reduce the degree of saturation of the concrete such as impermeable membranes.
2. Use of a low-alkali (less than 0.6% equivalent Na_2O) Portland cement; such a cement is available under BS 4027 (specification for sulphate-resisting Portland cement).
3. Limit the alkali content of the mix to a 3.0 kg/m³ of Na_2O equivalent.
4. Use of a ggbfs or pfa as composite cements or replacement materials in order that at least 50% ggbfs or 30% pfa by mass of the combined material are introduced into the mix.

One of the measures which can be used to minimize the risk of ASR is to limit the alkali content of the concrete mix to 3 kg/m³ of sodium oxide (Na_2O) equivalent. Currently a typical value for UK cements is 0.6–0.7%, say 0.65. It is important that this information should be obtained from the cement manufacturer on a regular basis. The alkali content of concrete can be expressed in the form:

$$A = (C \times a)/100$$
where
A = alkali content of concrete (kg/m³)
C = target mean Portland cement content of concrete (kg/m³)
a = equivalent sodium oxide in concrete (%)

For example, if $a = 0.65$ and the cement content is 350 kg/m³ and to allow for variation increase 'a' by 0.1 and 'C' by 10 kg/m³, then

$$A = (360 \times 0.75)/100$$
$$= 2.7 \text{ kg/m}^3$$

This result is satisfactory, and thus no further action need be taken other than a regular check on the weekly average alkali percentage as sodium equivalent.

Resistance to alkali–silica reaction (ASR) is covered in clause 5.7 of ENV 206. To minimize the risk of cracking or disruption of the concrete, one or more of the following precautions should be taken:

- limit the total alkali content of the concrete mix;
- use a cement with a low effective alkali content;
- change the aggregates;
- limit the degree of saturation of the concrete, e.g. by impermeable membranes.

Thus the precautions to be taken to minimize the risk of ASR listed in BS8110 and ENV206 are similar, but those in BS8110 are more explicit.

2.10 SUMMARY

Over 30 years ago, the explanatory handbook to CP 114: 1957 (Scott *et al.*, 1957) emphasized the great importance of ensuring that all reinforcement, particularly in members exposed to the weather, is protected by an adequate cover of well-compacted concrete, since experience has shown that corrosion of reinforcement and consequent spalling of the concrete have frequently resulted from inadequate cover. It is imperative that at the initial design stage attention is paid to the 'four Cs' guidelines for durability, that is:

- Constituents of the mix (including composition of cement)
- Cover
- Compaction
- Curing.

These should be related to the particular environmental conditions and the findings of recent research covered in Cather (1992) and NCE (1993). A useful summary of European concreting practice is given in a BRE information paper IP6/93 (Marsh, 1993).

REFERENCES

BRE (1982) The durability of steel in concrete: Part 2, Diagnosis and assessment of corrosion cracked concrete, *BRE Digest* No. 264, BRE, Garston, Watford.

Browne R.D. (1986) Practical considerations in producing durable concrete, *Proc. Seminar, How to Make Today's Concrete Durable for Tomorrow*, ICE, May 1985, Thomas Telford, London.

Browne R.D. (1987) Surface coatings for reinforced concrete, practical experience in the testing of coatings, *Half-day Meeting on Improvement in the Durability of Reinforced Concrete by Additives and Coatings*, ICE, February.

Cather R. (1992) How to get better curing (details of some of the thoughts of a Concrete Society Study Group about curing, with the intention of generating further discussion on the role of curing and the need for further research effort), *Concrete*, September/October.

CEIB (1992) Comité Euro-International du Béton, *Durable Concrete Structures (Design Guide)*, Thomas Telford, London.

IHVE (undated) *IHVE Guide Book A, Design Data*, Institution of Heating and Ventilation Engineers, London.

Marsh B.K. (1993) *European Concreting Practice: A Summary*, BRE Information Paper IP6/93, March.

NCE (1993) *Strong Resistance (An International Research Programme has Thrown Light on a Problem which has Plagued Concrete Specifiers for Many Years)*, NCE.

Nishi T. (1962) Outline of the studies in Japan regarding the neutralisation of alkali/or carbonation of concrete, *RILEM Int. Symp. on Testing of Concrete*, Prague.

Parrott L.J. (1987) *A Review of Carbonation in Reinforced Concrete*, Review carried out by C & CA under a BRE contract, July.

Powers T.C., Copeland L.E. and Mann H.M. (1959) Capillary continuity or discontinuity in cement pastes, *J. Portland Cement Assoc., Res. Dev. Labs.*, **1**, No. 2 (May), 38–48.

Scott W.L., Glanville W. and Thomas F.G. (1957) *Explanatory Handbook on the BS Code of Practice for Reinforced Concrete*, No. 114.

Trevatha G.T. (1961) *The Earth's Problem Climates*, University of Wisconsin Press, Madison.

Wallbank E.J. (1989) *The Performance of Concrete in Bridges (A Summary of 200 Highway Bridges)*, HMSO, London, April.

LOAD ARRANGEMENTS AND ANALYSIS

3.1 INTRODUCTION

EC2 lists the following behavioural idealizations used for analysis: elastic behaviour, elastic behaviour with limited redistribution, plastic behaviour, models and non-linear behaviour (cl. 2.5.1.1. P(5)). In this text, behavioural idealizations will, in general, be limited to:

- **Elastic behaviour**, e.g. for analysis of frameworks and continuous beams at ultimate and serviceability limit states.
- **Elastic behaviour with limited redistribution**, e.g. for analysis of continuous beams at ultimate limit state.
- **Plastic analysis**, e.g. for the analysis of slabs at ultimate limit state.

3.2 LOAD CASES AND COMBINATIONS (CL. 2.5.1.2)

Principle P(1) states that:

for the relevant combinations of actions, sufficient load cases shall be considered to enable the critical design conditions to be established at all sections within the structure or part of the structure considered.

Application clause P(2) indicates that:

depending on the type of structure, its function or the method of construction, design may be carried out primarily for either the ultimate or the serviceability limit state. In many cases, provided that checks for one of these limit states have been carried out, checks for the other may be disposed with as compliance can be seen by experience.

Taken to the extreme, the design of even a simple continuous slab could involve consideration of loading, member analysis and section analysis for both ultimate and serviceability limit states. A rigorous design of a four-span continuous slab would involve the steps shown in Table 3.1.

In accordance with application rules P(3), P(4) and P(5), it is normally adequate to adopt simplified combinations of actions and load cases, as below:

P(4)   For continuous beams and slabs in buildings without cantilevers subjected to dominantly uniformly distributed loads, it will generally be sufficient to consider only the following load cases (ULS):

(a) Alternate spans carrying the design variable and permanent loads ($\gamma_Q Q_k + \gamma_G G_k$). Other spans carrying only the design permanent load $\gamma_G G_k$.

(b) Any two adjacent spans carrying the design variable and permanent loads ($\gamma_Q Q_k + \gamma_G G_k$). All other spans carrying only the design permanent load $\gamma_G G_k$.

P(5)   For linear elements and slabs in buildings, the effects of shear and longitudinal forces on deformations may be ignored where these are likely to be less than 10% of those due to bending.

Table 3.1 Steps in design of four-span continuous slab

	ULS	*SLS*
(i) Loading	Six load cases	Six load cases
(ii) Member analysis	Evaluation of maximum bending moment and shear forces for load cases in (i)	Evaluation of maximum support and span moments for load cases in (i)
(iii) Section analysis	Evaluation of reinforcement requirements for (ii)	Evaluation of deflections and crack widths for (ii)

The effect of possible imperfections on the geometry of the unloaded structure will be considered in Chapter 5.

3.3 STRUCTURAL MODELS FOR OVERALL ANALYSIS (CL. 2.5.2.1)

The primary structural elements in conventional reinforced concrete frameworks are slabs, beams, columns and walls. Guidelines for the initial proportioning of these elements are given in Appendix A and they are defined in EC2 as follows:

To be considered as a beam or column, the span or length of the member should not be less than twice the overall section depth.

To be considered as a slab, the minimum span should not be less than four times the overall slab thickness. A slab subjected to dominantly uniformly distributed loads may be considered to be one-way spanning if either:

(a) it possesses two free (unsupported) and sensibly parallel edges, or

(b) it is the central part of a sensibly rectangular slab supported on four edges with a ratio of longer to shorter span greater than 2.

Ribbed or waffle slabs may be treated as solid slabs for the purposes of analysis provided there is sufficient stiffness. This may be assumed provided that the geometrical limitations shown in Figure 3.1 are complied with.

A wall should have a horizontal length of at least four times its thickness. Otherwise it should be treated as a column.

EC2 recommendations for determining effective widths of flanges in T/L beams (cl. 2.5.2.2.1) and effective span of beams and slabs (cl. 2.5.2.2.2) are summarized in Figure 3.2.

3.4 CALCULATION METHODS (CL. 2.5.3)

It is suggested that the following approach will simplify the calculation procedure and is in the spirit of EC2.

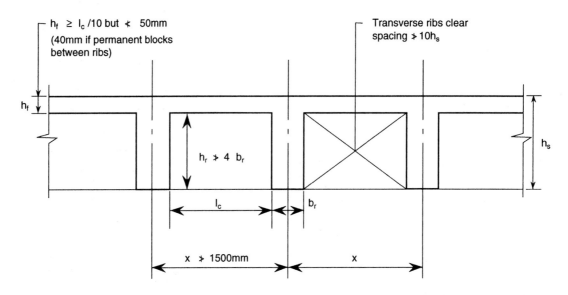

$h_f \geq l_c /10$ but $\not<$ 50mm
(40mm if permanent blocks between ribs)

Transverse ribs clear spacing $\not>$ $10h_s$

$h_r \not> 4\, b_r$

l_c

b_r

$x \not> 1500mm$

x

h_f

h_s

Figure 3.1 Requirements for ribbed or waffle slabs to be treated as solid slabs for analytical purposes.

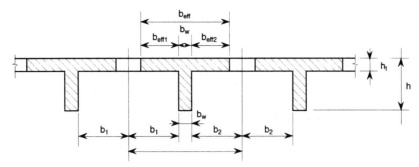

For analysis when a great accuracy is not required, for example, continuous beams in buildings a constant effective width (b_{eff}) may be assumed over the whole span. The effective width for a symmetrical T beam may be taken as

$$b_{eff} = b_w + \frac{1}{5} l_o \leq b$$

and for an edge beam, that is with floor on one side only

$$b_{effi} = b_w + \frac{1}{10} l_o \leq b_i + b_w \quad (i = 1 \text{ or } 2)$$

The distance l_o between points of zero moment may be obtained from the figure below for typical cases:

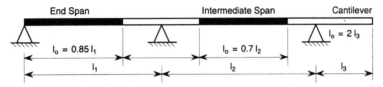

The following conditions should be satisfied:
(i)   The length of the cantilever should be less than half the adjacent span
(ii)   the ratio of adjacent spans should lie between 1 and 1.5

The effective span (l_{eff}) may be calculated as follows

$$l_{eff} = l_n + a_1 + a_2$$

where l_n is the clear distance between the faces of the supports and a_1 and a_2 are as in the figure below

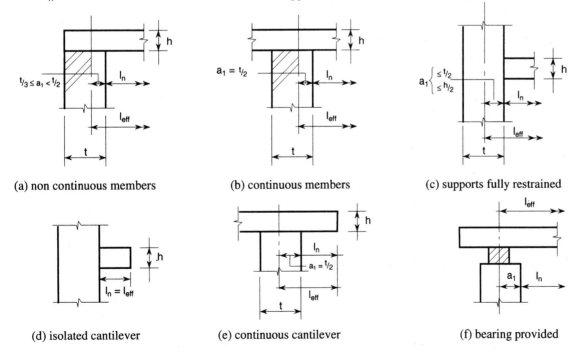

(a) non continuous members   (b) continuous members   (c) supports fully restrained

(d) isolated cantilever   (e) continuous cantilever   (f) bearing provided

Figure 3.2   Geometrical data for overall analysis.

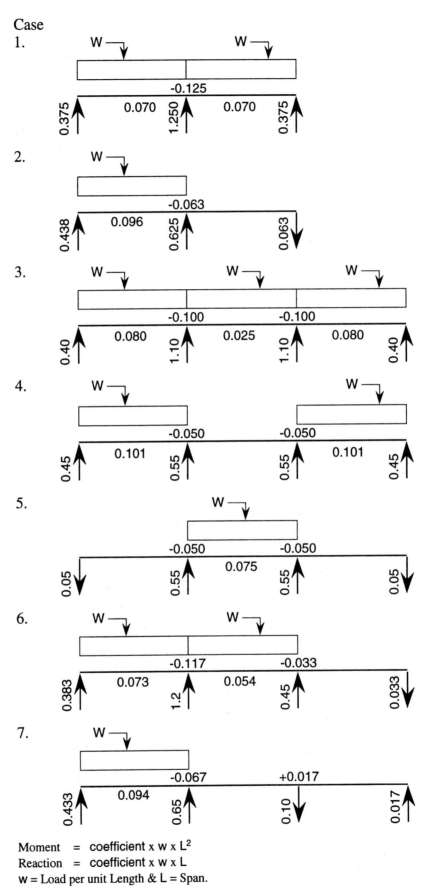

$$\text{Moment} = \text{coefficient} \times w \times L^2$$
$$\text{Reaction} = \text{coefficient} \times w \times L$$
$$w = \text{Load per unit Length} \ \& \ L = \text{Span}.$$

Figure 3.3 Equal-span continuous beams with uniformly distributed loads – elastic analysis (load cases 1–13).

Case

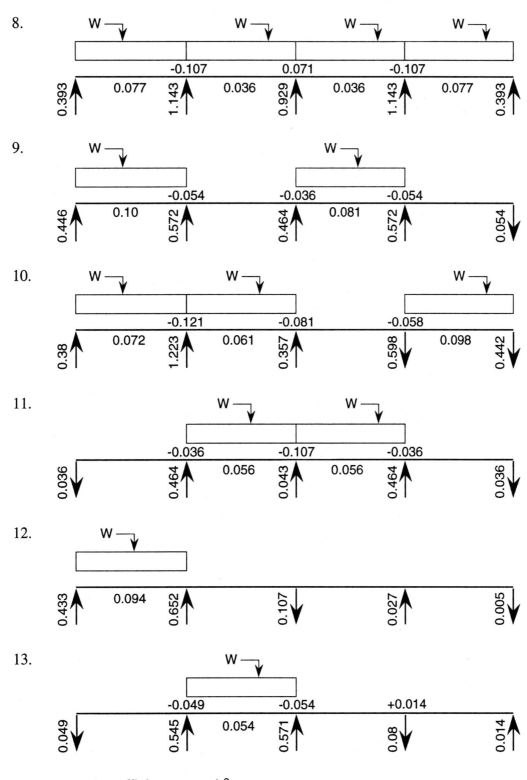

Moment = coefficient x w x L^2
Reaction = coefficient x w x L
w = Load per unit Length & L = Span.

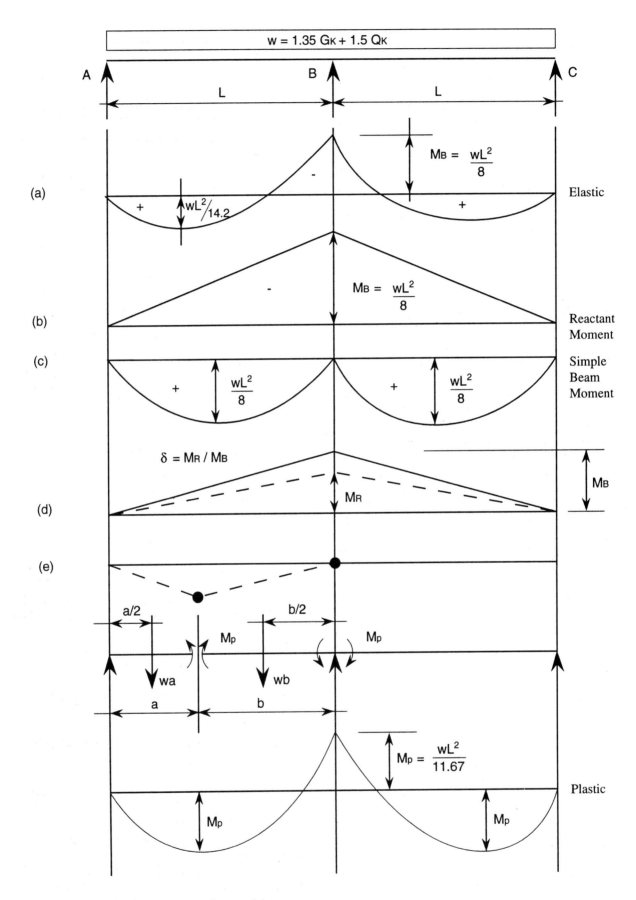

Figure 3.4   Analysis of two-span continuous slab.

3.4.1 Slabs

(a) Ultimate limit state

One of the following may be adopted:

1. The use of bending moment coefficients (see Figure 3.3) based on linear elastic theory, without redistribution.
2. As above, but with redistribution.
3. Plastic analysis using the yield-line theory (kinematic method). The use of the strip method (Hillerborg) is not in general appropriate for floor slabs in building frameworks but is a useful approach for rectangular tank walls and retaining walls. An example of the application of the strip method is given in Appendix E.

The three approaches to the analysis of slabs at ULS will be related to a simple example of a two-span continuous slab (see Figure 3.4). The moment at support B for an elastic analysis with a UDL of $w(1.35G_k + 1.5Q_k)$ per metre run on both spans is given by $M_{B(EL)} = wL^2/8$ and the corresponding span moment is $wL^2/14.2$. The elastic bending moment diagram is made up of the reactant diagram (b) and the simple beam diagram (c) where (a) = (b) + (c). If the assumption is made that the moment/rotation capacity is bilinear, the height of the reactant diagram at B can be adjusted; see (d). EC2 defines the ratio of the redistributed moment M_R to the moment before redistribution as δ, that is $\delta = M_R/M_{B(EL)}$. The value of δ, a function of the rotation capacity, is related to the neutral axis depth x, the concrete grade and the ductility of the steel as follows (cl. 2.5.3.4.1):

• For concrete grades not greater than C35/45

$$\delta \gtrsim 0.44 + 1.25x/d$$

which gives values of δ as below:

x/d	δ
0.1	0.57
0.2	0.69
0.25	0.75
0.3	0.82
0.35	0.88
0.4	0.94
0.45	1.00

• For concrete grades greater than C35/45

$$\delta \gtrsim 0.56 + 1.25x/d$$

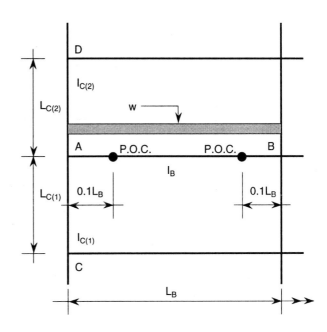

$$M_A = 0.4wL_B \times 0.1L_B + 0.1wL_B \times 0.05L_B$$
$$= 0.045wL_B{}^2$$

Figure 3.5 Point of contraflexure (POC) assumed at $0.1L_B$ to estimate beam end moments.

which gives value of δ as below:

x/d	δ
0.1	0.69
0.2	0.81
0.25	0.87
0.3	0.94
0.35	1.00

A further limitation on δ is that for high-ductility steel (see Chapter 1) $\delta \geq 0.7$, and for normal-ductility steel $\delta \geq 0.85$.

In elements where no redistribution is carried out, the ratio x/d should not exceed:

$x/d = 0.45$ for concrete grades C12/15 to C35/45

$x/d = 0.35$ for concrete grades C40/45 and greater

at the critical sections unless special detailing provisions are made (closely spaced links and the addition of longitudinal bars in the compression zone at critical sections can enhance rotation capacity; see Chapter 4).

A more direct approach for slabs is to adopt the yield-line theory (plastic analysis). This involves making an assumption with regard to the ratio of support to span moments. EC2 states that these should be between $\boxed{0.5}$ and $\boxed{2.0}$ and the corresponding range in the NAD is between 1.0 and 2.0. Assuming development of plastic hinges as in Figure 3.4(e) with equal support and span moments, then from considerations of equilibrium

$$M_p = wa^2/2$$

$$2M_p = wb^2/2$$

thus

$$1/2 = a^2/b^2$$

$$b = 1.414a$$

and

$$L = a + b = 2.414a$$

$$a = 0.414L$$

and

$$M_p = w(0.414L)^2/2 = wL^2/11.67$$

Assuming $\delta = 0.7$ at support B, then

$$M_{RB} = 0.7 \times wL^2/8 = wL^2/11.43$$

Thus use of $\delta = 0.7$ (the upper limit for high-ductility steel) gives comparable values of support and span moments to a plastic analysis with $M_{support}/M_{span}$ equal to unity:

	$M_{support}$	M_{span}
Plastic analysis	$wL^2/11.67$	$wL^2/11.67$
Redistribution ($\delta = 0.7$)	$wL^2/11.43$	$wL^2/11.75$

When using plastic methods for the analysis of slabs, EC2 limits the area of tensile reinforcement at any point, or in any direction, to a value corresponding to $x/d = 0.25$. A check on rotational capacity is not necessary for high-ductility steel. The limits imposed on the ratio of support to span moments are intended to ensure that there is not too great a departure from the ratio obtained from an elastic analysis. A large departure may produce problems with the serviceability analysis to check that control of deflection and crack widths is adequate. Reference should be made to the worked examples in Chapter 4.

It should be recognized that the object of a yield-line analysis (kinematic method) is to postulate a yield-line pattern (mechanism) from which the ultimate load of the slab can be evaluated for a given arrangement of reinforcement. Thus a variety of possible mechanisms should be examined to obtain the minimum value of the ultimate load for a given arrangement of reinforcement. Some standard cases are given in Appendix C.

(b) Serviceability limit state

It will not, in general, be necessary to undertake a detailed analysis in connection with serviceability limits, but, when necessary, a linear elastic analysis should be adopted (see example in Chapter 4).

3.4.2 Beams

Generally, the x/d ratio for beams will be greater than that for slabs and frequently in excess of 0.25. Thus a plastic analysis of beams will involve checking permissible and actual hinge rotations and the provision of confining reinforcement. This is not a practicable approach for beams in conventional reinforced concrete frameworks, unless seismic actions are dominant (see Chapter 7).

Thus member analysis for beams should be based on linear elastic analysis with or without redistribution, taking account of the restrictions in the ratio x/d if redistribution is adopted.

For the analysis of beams and slabs, EC2 gives a number of simplifications (cl. 2.5.3.3) as below:

P(1) Simplified methods or design aids based on appropriate simplifications may be used for analysis provided they have been formulated to give the level of reliability implicit in the methods given in this code over their stated field of validity. Redistribution is limited to that permitted by the assumptions implicit in the chosen simplified method.

P(2) A value of zero may be taken for Poisson's ratio.

P(3) Continuous slabs and beams may generally be analysed on the assumption that the supports provide no rotational restraint.

P(4) Regardless of the method of analysis used, where a beam or slab is continuous over a support that may be considered to provide no restraint to rotation, the design support moment, calculated on the basis of a span equal to the centre-to-centre distance between supports, may be reduced by an amount δM_{Sd} given by

$$\delta M_{Sd} = F_{Sd, \, sup} \, b_{sup}/8$$

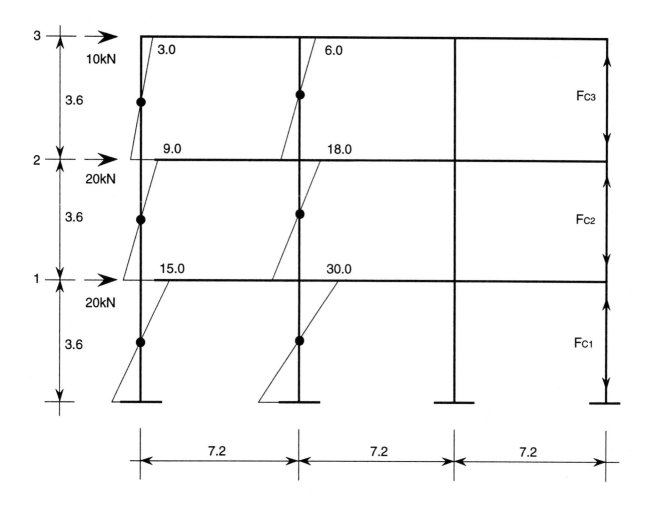

STOREY	TOTAL WIND SHEAR (kN)	SHEAR FORCE IN COLUMN (kN)		BENDING MOMENT IN COLUMN (kN)	
		EXTERNAL	INTERNAL	EXTERNAL	INTERNAL
3	10	1.67	3.33	1.67 x 1.8 = 3.01	3.33 x 1.8 = 6.0
2	30	5.0	10.0	5.0 x 1.8 = 9.0	10.0 x 1.8 = 18.0
1	50	8.33	16.66	8.33 x 1.8 = 15.0	16.66 x 1.8 = 30.0

STOREY	MOMENTS ABOUT P. DFC. AT MID-HEIGHT		AXIAL LOAD IN COLUMN (kN)
3	$21.6 F_{C3}$ =	10.0×1.8	F_{C3} = 0.83
2	$21.6 F_{C2}$ =	$10.0 \times 5.4 + 20.0 \times 1.8$	F_{C2} = 4.17
1	$21.6 F_{C1}$ =	$10.0 \times 7.2 + 20.0 \times 5.4 + 20.0 \times 1.8$	F_{C1} = 10.0

Figure 3.6 Use of 'portal' method to determine column moments and axial loads for 10 kN, 20 kN and 20 kN at levels 3, 2 and 1 respectively.

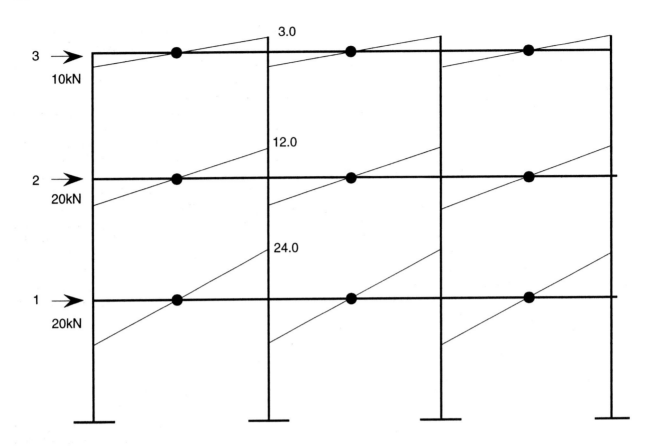

Figure 3.7 Use of 'portal' method to determine beam moments.

FLOOR LEVEL	BENDING MOMENT IN EXTERNAL COLUMN (kN)		BENDING MOMENT IN BEAM (kNm)
	UPPER COLUMN	LOWER COLUMN	
3	-	3.0	0 + 3.0 = 3.0
2	3.0	9.0	3.0 + 9.0 = 12.0
1	9.0	15.0	9.0 + 15.0 = 24.0

Where $F_{Sd, sup}$ is the design support reaction and b_{sup} is the breadth of the support.

P(5) Where a beam or slab is cast monolithically into its supports, the critical design moment at the support may be taken as that at the face of the support (but not less than 65% of the support moment calculated using full fixity at the faces of the rigid supports – cl. 2.5.3.4.2 P(7)).

P(6) The loads applied to supporting members by the reactions from one-way spanning slabs, ribbed slabs and beams (including T beams) may be calculated on the assumption that the members supported are simply supported. Continuity should, however, be taken into account at the first internal support and at other internal supports if the spans on either side of the support differ by more than 30%.

3.4.3 Frames

An initial estimate of the element proportions, slabs, beams and columns can be obtained from the guidelines given in Appendix B; thus G_k can be determined. For hand analysis the following procedure is suggested.

(a) Vertical loads

The beams may be analysed on the assumption that the supports provide no rotational restraint using the bending moment coefficients in Figure 3.3. Column moments arising from vertical loads may be estimated by assuming a point of contraflexure at $0.1L_B$, see Figure 3.5. Considering joint A,

$$M_A = 0.4 \, wL_B \times 0.1L_B + 0.1 \, wL_B \times 0.05 \, L_B$$

$$= 0.045 \, wL_B^2$$

The moment in the upper column AD is

$$M_{AD} = M_A \frac{I_{C(2)}/L_{C(2)}}{I_B/L_B + I_{C(2)}/L_{C(2)} + I_{C(1)}/L_{C(1)}}$$

and that in the lower column AC is

$$M_{AC} = M_A \frac{I_{C(1)}/L_{C(1)}}{I_B/L_B + I_{C(2)}/L_{C(2)} + I_{C(1)}/L_{C(1)}}$$

I_B (normally a T section) is obtained from the geometrical data given in Figure 3.2.

(b) Horizontal loads

Moments and shears arising from horizontal loads may be estimated for preliminary design using the 'portal' method (SCI, 1991) (see Figures 3.6 and 3.7). The total horizontal (e.g. wind) shear is divided between the bays in proportion to their spans. For an internal column, the shear is obtained by summing the contributions to the column from adjacent bays. The internal moment is obtained by multiplying the shear by half the storey height (the assumed distance from the point of contraflexure to the end of the column). The axial force in an external column is obtained by taking moments about the point of contraflexure, for the portion of the frame above the level being considered. Forces are compressive in the leeward column, tensile in the windward column. The axial force in an internal column, due to wind, is zero.

With the general availability of standard computer packages for linear analysis of plane frameworks (e.g. SAND), it will normally be more convenient to adopt a computer analysis for final design. This has the advantage of outputting horizontal displacements, which will enable inter-storey drift and P/δ effects to be evaluated; see Chapters 6 and 7.

A hand analysis using the approach outlined above may also be used for checking computer output. CIRIA Technical Note 133 (CIRIA, 1988) gives useful guidelines for checking computer analysis of building structures. Chapter 8 provides detailed calculations for a multi-storey framework using manual methods.

3.5 SUMMARY

The following procedure is suggested.

- **Slabs** – For the ultimate limit state with approximately equal spans and uniformly distributed loads, adopt elastic bending moment coefficients (with or without redistribution) or yield-line theory, noting restrictions in x/d ratios.
- **Beams** – For the ultimate limit state with approximately equal spans and uniformly distributed loads, adopt elastic bending moment coefficients (with or without redistribution), noting restrictions in x/d ratios.
- **Frames** – For the ultimate limit state, adopt the 'portal' method for preliminary design and computer analysis for final design.

A linear elastic analysis should be adopted for slabs, beams and frames (when necessary) for serviceability checks.

REFERENCES

CIRIA (1988) *Guidelines for Checking Computer Analysis of Building Structures*, CIRIA Technical Note 133, Construction Industry Research and Information Association.

SCI (1991) *Wind Moment Analysis for Unbraced Frames*, SCI Publ. 082, Steel Construction Institute.

SECTION ANALYSIS (1): SLABS AND BEAMS

4.1 INTRODUCTION

Following the member analysis as outlined in Chapter 3, the next stage is to undertake a section analysis for the ultimate and serviceability limit states. This chapter deals with section analysis for slabs and beams; columns and walls are considered in Chapter 5 and aspects of joint design relevant to seismic actions (EC8 – Draft) in Chapter 6. The starting point is section analysis for the ultimate limit state (flexure) to obtain the reinforcement ratio (p_1) for longitudinal (tension) reinforcement. The reinforcement ratio (p_1) influences subsequent calculations for the ultimate limit state (shear and torsion) and those for the serviceability limit state (crack width and deflection control). A number of worked examples are included that link loading, section and member analysis for the two limit states. The interaction of EC8 (Draft) and EC2 vis-$à$-vis beam design and detailing is also covered.

4.2 ULTIMATE LIMIT STATE – FLEXURE (CL. 4.2.1.3)

In common with BS 8110, sections are proportioned for flexure to ensure an under-reinforced failure mode, that is, the steel yields before the concrete reaches its limiting strain (0.0035 for unconfined concrete). This is achieved by limiting the neutral axis depth (x). The design stress–strain relationships for concrete in compression and reinforcing steel are shown in Figure 4.1. In this chapter, a simplified rectangular stress block for concrete is assumed. The design charts given in Appendix G are based on a parabolic rectangular stress block. The equivalent rectangular stress blocks, EC2 and BS 8110, are shown in Figure 4.2 and equations developed for the moment of resistance of rectangular sections. The following differences between EC2 and BS 8110 should be noted:

1. EC2 uses cylinder strengths (f_{ck}) whereas BS 8110 uses cube strengths (f_{cu}). The f_{ck}/f_{cu} relationship is given at the top of Table 1.6.
2. EC2 introduces a reduction factor α for sustained compression, which may generally be assumed to be $\boxed{0.85}$ except when the compression zone decreases in width in the direction of the extreme compression fibres.
3. EC2 restricts the neutral axis depth as follows:

    $x = 0.45d$ for $f_{ck} \leqslant 35$ N/mm^2,
    no redistribution (for redistribution, see Chapter 3)

    $x = 0.35d$ for $f_{ck} > 35$ N/mm^2

    $x = 0.25d$ for plastic analysis

4. BS 8110 restricts the neutral axis depth to $x = 0.5d$ (no redistribution).
5. EC2 uses a compression zone depth of $0.8x$ whereas BS 8110 uses $0.9x$.

A design chart for flexure (EC2) can be obtained by plotting $M/b_w d^2$ against x/d for various values of f_{ck} (20–50 N/mm^2); see Figure 4.3. Note the upper limit of $x/d = 0.35$ for concrete grade C40 and greater. The chart relates to rectangular sections, but can be used for T/L beams if x is not greater than h_f. If x is greater than h_f, then the area of

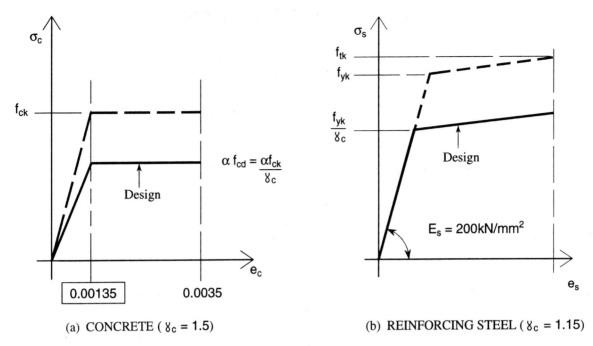

Figure 4.1 EC2 design stress–strain relationships: (a) concrete, simplified from parabolic rectangular curve; (b) reinforcing steel.

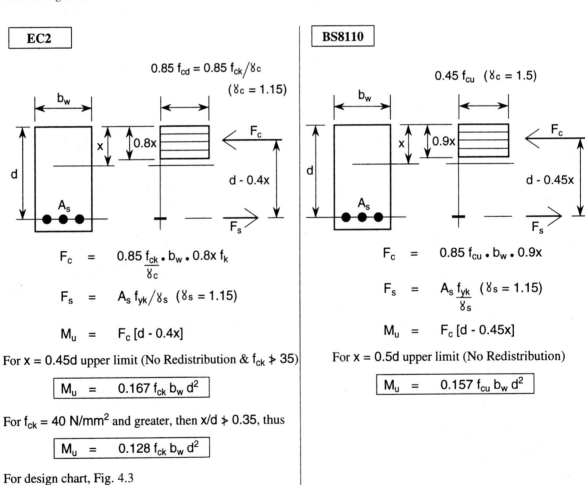

Figure 4.2 Simplified rectangular stress blocks for EC2 and BS 8110.

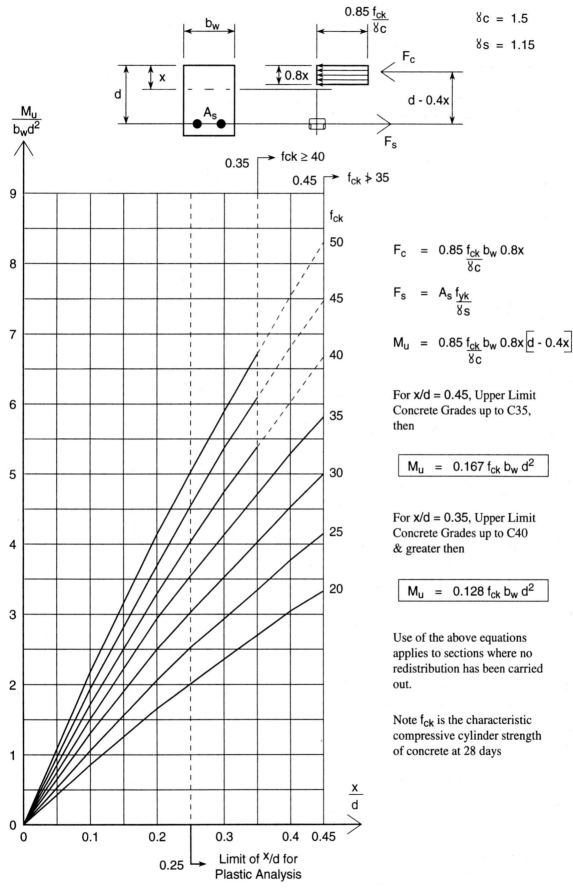

The chart shows the vertical axis $\frac{M_u}{b_w d^2}$ ranging from 0 to 9, and the horizontal axis $\frac{x}{d}$ ranging from 0 to 0.45.

$\gamma_c = 1.5$

$\gamma_s = 1.15$

$0.85 \frac{f_{ck}}{\gamma_c}$

F_c

$d - 0.4x$

F_s

0.85

0.35 → fck ≥ 40

0.45 → fck ≱ 35

$F_c = \dfrac{0.85\, f_{ck}\, b_w\, 0.8x}{\gamma_c}$

$F_s = \dfrac{A_s\, f_{yk}}{\gamma_s}$

$M_u = \dfrac{0.85\, f_{ck}\, b_w\, 0.8x}{\gamma_c}\left[d - 0.4x\right]$

For x/d = 0.45, Upper Limit Concrete Grades up to C35, then

$$M_u = 0.167\, f_{ck}\, b_w\, d^2$$

For x/d = 0.35, Upper Limit Concrete Grades up to C40 & greater then

$$M_u = 0.128\, f_{ck}\, b_w\, d^2$$

Use of the above equations applies to sections where no redistribution has been carried out.

Note f_{ck} is the characteristic compressive cylinder strength of concrete at 28 days

0.25 → Limit of x/d for Plastic Analysis

Figure 4.3 EC2 design chart for flexure.

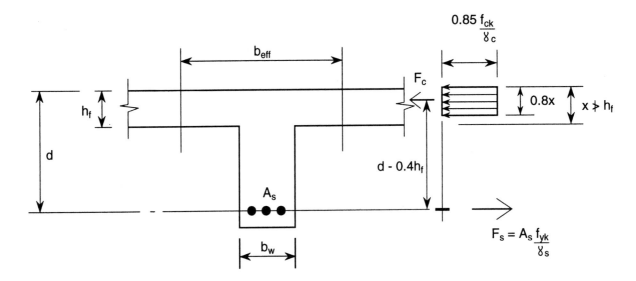

Figure 4.4 Simplified procedure for T beams.

concrete in compression in the web may be conservatively ignored and the ultimate moment of resistance of the T/L beam can be expressed, see Figure 4.4, as:

$$M_u = 0.85(f_{ck}/\gamma_c)b_{eff} \times 0.8h_f(d - 0.4h_f)$$

and with $\gamma_c = 1.5$,

$$M_u = 0.453f_{ck}b_{eff}h_f(d - 0.4h_f) \qquad (4.1)$$

The use of the design chart (Figure 4.3) enables the value of x/d to be immediately established and thus the suitability of the member for plastic analysis or moment redistribution. Maximum values of the reinforcement ratio $p_1 = A_s/b_wd$ for longitudinal tension reinforcement are given in Table 4.1. These correspond to the maximum value of x/d for a range of concrete cylinder strengths (f_{ck}) and characteristic yield strengths (f_{yk}) of the reinforcement. Minimum reinforcement percentages will be covered in the

section on detailing requirements at the end of this chapter.

4.3 ULTIMATE LIMIT STATE – SHEAR (CL. 4.3.2.2)

In order to take full advantage of the potential ductility of reinforced concrete, it is desirable to ensure that ultimate strengths are governed by flexure rather than shear. The analysis of reinforced concrete sections under the combined action of bending and shear is a complex problem. The forces acting on a reinforced concrete section (without shear reinforcement) are shown in Figure 4.5. The factors contributing to shear resistance are:

- **compression zone shear**, V_1 (20–40%);
- **mechanical interlock** of the aggregate at the crack, V_2 (35–50%);
- **dowel action** of the longitudinal reinforcement, V_3 (15–25%).

Table 4.1 Maximum values of $p_1 = A_s/b_wd$ related to f_{ck} and f_{yk} (note $x \leq 0.35d$ for $f_{ck} > 35$ N/mm², and $x \leq 0.45d$ for $f_{ck} \leq 35$ N/mm²)

f_{ck} (N/mm²)	f_{yk} (N/mm²)				Maximum x/d, no redistribution
	250	*400*	*460*	*500*	
20	0.019	0.012	0.010	0.009	0.45
25	0.024	0.015	0.013	0.012	0.45
30	0.028	0.018	0.015	0.014	0.45
35	0.033	0.021	0.018	0.016	0.45
40	0.029	0.018	0.016	0.015	0.35
45	0.033	0.021	0.018	0.016	0.35
50	0.036	0.023	0.020	0.018	0.35

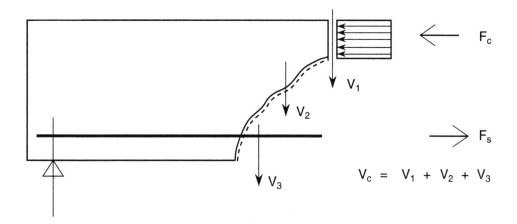

Figure 4.5 Factors contributing to shear resistance.

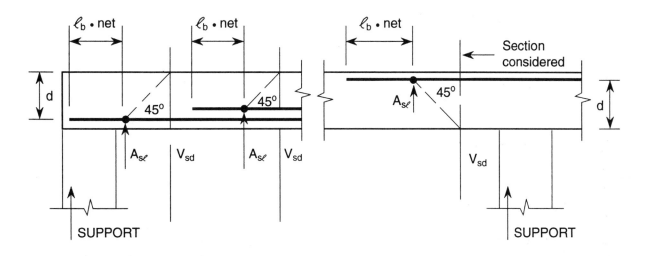

Figure 4.6 Evaluations of A_{sl} for use in equations for V_{Rd1}.

Attempts have been made (Kong and Evans, 1987) to quantify the relative magnitude of V_1, V_2 and V_3, and the figures in brackets above are for a typical reinforced concrete beam. The shear resistance of the concrete V_c can be expressed as:

$$V_c = V_1 + V_2 + V_3$$

The shear resistance of the concrete is influenced by the tension steel ratio ($p_1 = A_s/b_w d$), dowel action, beam depth and aggregate interlock, and these are reflected in the design clauses for shear in both BS 8110 and EC2. For slabs, the shear capacity of the concrete (with the exception of flat slabs) will generally be adequate, but for beams, the provision of shear reinforcement in the form of stirrups (links) and/or bent-up bars is the norm. The provision of shear reinforcement will enhance the ductility of the beam – particularly important in seismic design –

and reduce the probability of sudden, brittle failure. The total shear capacity (V_T) may be expressed in the form:

$$V_T = V_1 + V_2 + V_3 + V_s$$
$$= V_c + V_s$$

where V_s is the contribution of the shear reinforcement. In EC2 the design method for shear is based on the generalized expression $V_T = V_c + V_s$, which is based on three values of design shear resistance: V_{Rd1}, V_{Rd2} and V_{Rd3}.

(i) V_{Rd1} is the design shear resistance of the element without shear reinforcement. Thus the design value of the applied shear force (V_{sd}) must be less than V_{Rd1}. Ignoring the presence of axial forces, the design shear resistance V_{Rd1} may be expressed as:

Table 4.2 Values of V_{Rd2} (applied axial compression assumed to be negligible)

f_{ck}	$f_{cd} = f_{ck}/\gamma_c$	$f_{ck}/200$	$v = 0.7 - f_{ck}/200$	$V_{Rd2} = 0.45vf_{cd}\,b_w d$
20	13.33	0.1	0.6	$3.6b_w d$
25	16.66	0.125	0.575	$4.31b_w d$
30	20.00	0.15	0.55	$4.95b_w d$
35	23.33	0.175	0.525	$5.51b_w d$
40	26.66	0.20	0.5	$6.0b_w d$
45	30.00	0.225	0.5	$6.75b_w d$
50	33.33	0.250	0.5	$7.5b_w d$

$$V_{Rd1} = \tau_{Rd}k(1.2 + 40p_1)b_w d \qquad (4.2)$$

Here τ_{Rd} = basic design shear strength of concrete = $0.035f_{ck}^{2/3}$ (N/mm²); for tabulated values see Table 1.6. This includes $\gamma_c = 1.5$. The value of $k = \boxed{1}$ for members where more than 50% of the bottom reinforcement is curtailed, otherwise $k = \boxed{1.6 - d \geqslant 1}$ (d in metres); $p_1 = A_{sl}/b_w d \leqslant \boxed{0.02}$ and A_{sl} = the area of tension reinforcement extending not less than $d + l_{b,net}$ beyond the section considered, see Figure 4.6; $l_{b,net}$ is the required anchorage length and is quantified in the section on bond; b_w = minimum width of section over the effective depth.

Thus the expression for V_{Rd1} recognizes the influence of dowel action and member depth. If V_{sd} is greater than V_{Rd1} then shear reinforcement must be provided.

(ii) Irrespective of the amount of shear reinforcement provided, an upper limit is imposed on the design shear resistance to preclude web crushing or compression zone failure. The design resistance V_{Rd2} is given by:

$$V_{Rd}2 = 1/2vf_{cd}b_w \times 0.9d \qquad (4.3)$$

where v (efficiency factor) is

$$v = 0.7 - f_{ck}/200 \geqslant 0.5 \qquad (f_{ck} \text{ in N/mm}^2)$$

Values of V_{Rd2} in the range $f_{ck} = 20$ to 50 N/mm² are given in Table 4.2. V_{Rd2} is roughly equivalent to the BS 8110 expression $0.8f_{cu}^{1/2}b_w d$, again putting an upper limit on shear resistance irrespective of the shear reinforcement provided.

(iii) V_{Rd3} is the design shear that can be carried by a member with shear reinforcement. V_{Rd3} is expressed as:

$$V_{Rd3} = V_{Rd1} + V_{wd} \qquad (4.4)$$

where V_{wd} is the contribution of the shear reinforcement. EC2 gives two methods of proportioning

shear reinforcement – the standard method and the variable strut inclination method. In the interests of simplicity, the standard method only will be considered in relation to vertical stirrups. The contribution of the vertical stirrups is given by the expression:

$$V_{wd} = (A_{sw}/s) \times 0.9df_{ywd} \qquad (4.5)$$

where A_{sw} is the cross-sectional area of the vertical links and s is the spacing of the stirrups. EC2 requirements for flexure and shear are brought together in the following example. Detailing provisions are considered at the end of the chapter.

Example 4.1: flexure and shear capacity

Given the following data, estimate the maximum flexural capacity (without redistribution) and the maximum shear capacity (with and without shear reinforcement) at the ultimate limit state:

$d = 600$ mm $f_{ck} = 30$ N/mm²

$b_w = 300$ mm $f_{yk} = 460$ N/mm²
(tension reinforcement and stirrups)

The maximum flexural capacity is given by:

$$M_u = 0.167f_{ck}b_w d^2 \qquad (f_{ck} \not> 35 \text{ N/mm}^2)$$

$$= 0.167 \times 30 \times 300 \times 600^2 \times 10^{-6}$$

$$= 541.08 \text{ kNm}$$

The corresponding reinforcement ratio (p_1) from Table 4.1 is 0.015 with $x/d = 0.45$.

The basic shear strength is

$$\tau_{Rd} = 0.035 \times 30^{2/3} = 0.34 \text{ N/mm}^2$$

$$V_{Rd1} = \tau_{Rd}k(1.2 + 40 p_1) b_w d$$

Loaded area shown hatched.

d is the average effective depth of slab for circular columns, diameter ⊁ |3.5d|.

For rectangular columns, perimeter ⊁ |11d| & ratio of length to breadth ⊁ |2|.

(a) & (b) Critical perimeter away from unsupported edge

(c) & (d) Critical section near unsupported edge

(e) Critical perimeter near an opening

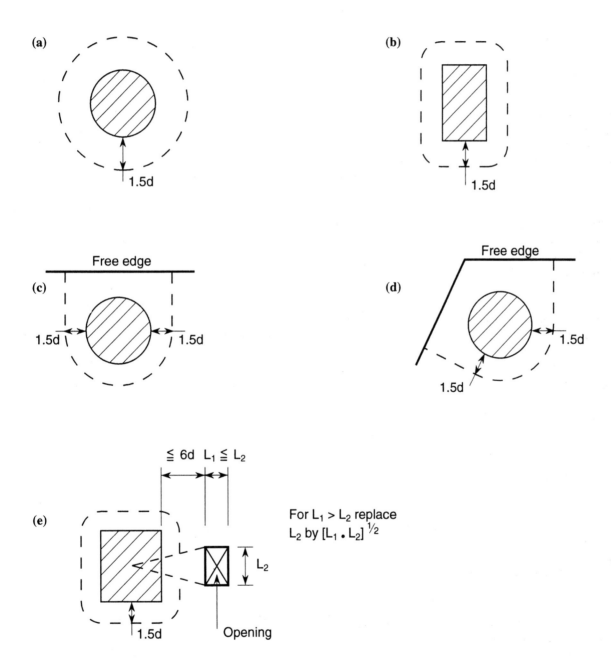

Figure 4.7 Critical perimeters for flat slabs, standard cases. For non-standard cases and foundations, refer to EC2 cl. 4.3.4.1/2.

$$k = 1.6 - 0.6 = 1.0$$

Thus

$$V_{Rd1} = 0.34 \times 1.0(1.2 + 40 \times 0.015)300$$
$$\times 600 \times 10^{-3}$$
$$= 110.16 \text{ kN}$$

The maximum shear capacity V_{Rd2} (see Table 4.2) is

$$V_{Rd2} = 4.95b_w d$$
$$= 4.95 \times 300 \times 600 \times 10^{-3}$$
$$= 891 \text{ kN}$$

Thus the shear reinforcement V_{wd} required is

$$V_{wd} = 891 - 110.16 = 780.84 \text{ kN}$$

Assuming 12 m dia. stirrup (four legs), $A_{sw} = 452$ mm^2,

$$s = A_{sw} f_{ywd} \times 0.9d/V_{wd}$$
$$= 452 \times 460 \times 0.9 \times 600/780.84 \times 10^3 \times 1.15$$
$$= 125 \text{ mm}$$

Note that in the NAD, the maximum value of f_{ck} in the determination of τ_{Rd} should be taken as 40 N/mm^2 (strength class C40/50).

4.4 PUNCHING SHEAR (CL. 4.3.4.3)

As in section 4.3, design for punching shear is based on three values of design shear resistance, that is:

- V_{Rd1}, the design shear resistance per unit length of the critical perimeter, for a slab without shear reinforcement.
- V_{Rd2}, the maximum design shear resistance per unit length of the critical perimeter, for a slab with shear reinforcement.
- V_{Rd3}, the design shear resistance per unit length of the critical perimeter, for a slab with shear reinforcement.

If the design shear force (V_{sd}) is less than V_{Rd1}, then no shear reinforcement is required. If $V_{sd} > V_{Rd1}$, then shear reinforcement, if appropriate, should be provided such that $V_{sd} > V_{Rd3}$. Noting that EC2 refers to shear per unit length, then in the case of a concentrated load or support reaction, the applied shear per unit length is given by

$$V_{sd} = V_{sd} \beta/u \qquad (4.6)$$

where u is the perimeter of the critical section, which can be obtained from Figure 4.7, and β is a coefficient that takes into account the effects of any eccentricity of loading. In the absence of a rigorous analysis, values of $\beta = 1.5$, 1.4 and 1.15 may be used for corner, edge and internal columns respectively.

In a slab, the total design shear force developed (V_{sd}) is calculated along the perimeter u. The shear resistance values V_{Rd1}, V_{Rd2} and V_{Rd3} are obtained as below. Only flat slabs of constant depth are considered and reference should be made to EC2 (cl. 4.3.4.4) for slabs with variable depth and slabs with column heads.

(i) V_{Rd1} (cl. 4.3.4.5.1) – The shear resistance per unit length of slabs and foundations (non-prestressed) is obtained by a similar formula to that given in section 4.3, that is

$$V_{Rd1} = \tau_{Rd}k(1.2 + 40p_1)d \qquad (4.7)$$

τ_{Rd} and k have been defined previously and p_1 relates to tension steel in the x and y directions respectively, p_{1x} and p_{1y},

$$p_1 = (p_{1x}p_{1y})^{1/2} \leqslant 0.015$$

$$d = (d_x + d_y)/2$$

where d_x and d_y are the effective depths of the slab at the points of intersection between the design failure surface and the longitudinal reinforcement in the x and y directions respectively.

(ii) V_{Rd2} (cl. 4.3.4.5.2) – Flat slabs containing shear reinforcement should have a minimum depth of 200 mm and the maximum shear resistance is given by

$$V_{Rd2} = \boxed{1.6} V_{Rd1}$$

In the NAD:

$$V_{Rd2} = 2.0V_{Rd1}$$

but the shear stress at the perimeter of the column should not exceed $0.9f_{ck}^{1/2}$.

(iii) V_{Rd3} (cl. 4.3.4.5.2) – The maximum shear reinforcement that can be provided is

$$0.6V_{Rd1} \text{ (EC2)} \quad \text{or} \quad 1.0V_{Rd1} \text{ (NAD)}$$

Where the shear reinforcement provided is greater than $0.6V_{Rd1}$ but not greater than $1.0V_{Rd1}$, reference should be made to clause 6.4(d) of the NAD.

Practical guidelines on shear reinforcement systems for flat slabs have recently been published by the British Cement Association (BCA, 1990).

To ensure that the punching shear resistance in (i), (ii) and (iii) above can be fully developed, the slab should be designed for minimum bending moments per unit width (cl. 4.3.4.5.3) in the x and y directions; see Table 4.9 of EC2.

Example 4.2: punching shear, flat slab

Consider a flat slab (Figure 4.8) supported by a column grid with $L_x = L_y = 6.0$ m. The overall slab depth is 200 mm and the columns are 400×400 mm^2. The reinforcement is 12φ–150 c/c (754 mm^2/m) both ways top and bottom and the cover is 20 mm. Given the following data, estimate adequacy of the slab to resisting punching at an internal column without shear reinforcement:

$$f_{ck} = 40 \text{ N/mm}^2 \qquad Q = 2.5 \text{ kN/m}^2$$

$$f_{yk} = 460 \text{ N/mm}^2 \qquad G = 0.2 \times 24 = 4.8 \text{ kN/m}^2$$

The design shear force, neglecting the load between the column perimeter and the critical perimeter, is

$$V_{sd} = (1.35 \times 4.8 + 1.5 \times 2.5)6^2 = 368.28 \text{ kN}$$

The critical perimeter (see Figure 4.8) is

$$u = 4 \times 400 + 2\pi \times 1.5d$$

From Figure 4.8

$$d_x = 200 - 26 = 174 \text{ mm}$$

$$d_y = 200 - 38 = 162 \text{ mm}$$

Average effective depth

$$d = (174 + 162)/2 = 168 \text{ mm}$$

Thus

$$u = 1600 + 6.28 \times 1.5 \times 168 = 3182.56 \text{ mm}$$

For an internal column, the applied shear per unit length is

$$\begin{aligned} V_{sd} &= V_{sd}\beta/u \qquad (\beta = 1.15) \\ &= 368.28 \times 10^3 \times 1.15/3182.56 \\ &= 133.1 \text{ N/mm} \end{aligned}$$

The shear resistance per unit length is given by

$$V_{Rd1} = \tau_{Rd}k(1.2 + 40p_1)d$$

for $f_{ck} = 40$ N/mm^2

$$\tau_{Rd} = 0.41$$

$$k = 1.6 - 0.168 = 1.432$$

$$p_{1x} = 754/10^3 \times 174 = 0.0043$$

$$p_{1y} = 754/10^3 \times 162 = 0.0047$$

$$p_1 = (p_{1x}p_{1y})1/2 = 0.0045$$

Thus

$$\begin{aligned} V_{Rd1} &= 0.41 \times 1.432 \ (1.2 + 40 \times 0.0045)168 \\ &= 136.1 \text{ N/mm} \end{aligned}$$

Thus the slab is just adequate without shear reinforcement or provision of a column head.

It is necessary to check that the minimum bending moments per unit width m_{sdx} and m_{sdy} have been provided and clause 4.3.4.5.3 of EC2 requires that

$$m_{sdx} \text{ (or } m_{sdy}) > nV_{sd}$$

From Table 4.9 of EC2, the value of n (internal column) for m_{sdx} and m_{sdy} is 0.125 for the top and zero for the bottom of the slab with an effective width of $0.3L_x$ or $0.3L_y$. In this example

$$\begin{aligned} M_{sdx} &= M_{sdy} = 0.125 \times 368.28 \\ &= 46.035 \text{ kN m} \quad \text{over 1.8 m width } (0.3L_x) \end{aligned}$$

$$\begin{aligned} \text{force in reinforcement} &= A_s f_{yk}/\gamma_s \\ &= 754 \times 460/1.15 \\ &= 301\,600 \text{ N} \end{aligned}$$

$$\text{force in concrete} = (0.85f_{ck}/\gamma_c)b_w \times 0.8x$$

$$= 18133.3x$$

Thus

$$x = 16.63 \text{ mm}$$

$$d - 0.4x = 162 - 0.4 \times 16.63 = 155.35$$

$$M_u = 301\,600 \times 155.35 \times 10^{-6} = 46.85 \text{ kN m/m}$$

Thus reinforcement provided (754 mm^2/m) is adequate to meet the minimum requirements. This

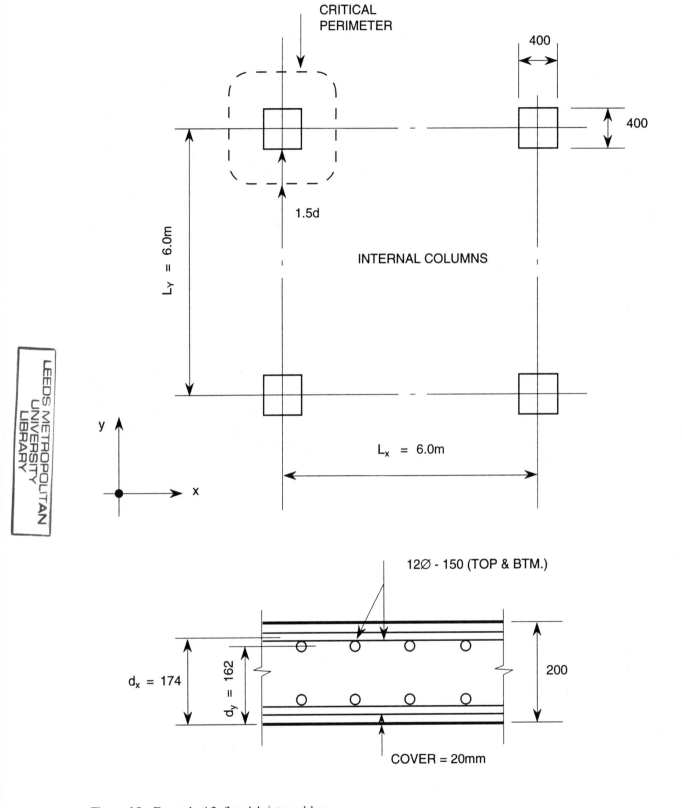

Figure 4.8   Example 4.2: flat slab internal bay.

example demonstrates the limitations of a flat slab without shear reinforcement and column heads. For column grids greater than 6 m × 6 m and imposed loads other than residential, shear reinforcement and/or column heads will generally be required. For a detailed treatment of flat slabs, including basic equilibrium requirements, variations from simple flat plate construction, methods of analysis, punching shear and methods of reinforcement, consult BCA (1990) and Regan (1986).

4.5 TORSION (CL. 4.3.3)

EC2 requires a full design for torsion covering both the ultimate and serviceability limit states where the static equilibrium of the structure depends upon the torsional resistance of the elements of the structure, for example, curved and cranked beams. In conventional reinforced concrete frameworks it is generally possible to arrange the structural elements such that it will not be necessary to consider torsion at the ultimate limit state. However, torsions arise from consideration of compatibility only, which may lead to excessive cracking in the serviceability limit state. An example of this is shown in Figure 4.9. A long span slab is supported by deep edge beams, which in turn are eccentrically connected to columns. The edge beam will be subjected to combined flexure, shear and torsion, and the torsional moments in the edge beams will induce additional flexure in the columns. Thus excessive cracking can occur in both

the beam and column if the effects of torsion are ignored. The configuration of structural elements shown in Figure 4.9 is not uncommon in high-rise buildings and multi-storey car parks and should be avoided unless torsional actions are included in the design. Compatibility torsion is recognized in EC2 (cl. 4.3.3.1) and the need to limit excessive cracking noted.

The equations for torsional resistance of sections in BS 8110 are derived from the sand heap analogy, whereas in EC2, the thin-walled closed section forms the basis of the torsional resisting moment equations. Consider a thin-walled closed section (Figure 4.10), subjected to a torsional moment T. The shear flow q per unit length round the centre-lines of the sections walls provides the internal resistance to the applied torsion. Thus

$$T = 2(qhb/2 + qbh/2) = 2qbh \qquad (4.8)$$

The torsion shear stress $v_t = q/t$. Thus

$$T = 2v_t tbh \qquad (4.9)$$

bh represents the area enclosed by the centre-line of all the walls and is given the notation A_k in EC2. As with the shear clauses (see section 4.3), the concrete stress in the struts is limited to $\sigma_c < v f_{cd}$ where v is the efficiency factor. Thus T can be expressed in the form

$$T = 2v f_{cd} t A_k \qquad (4.10)$$

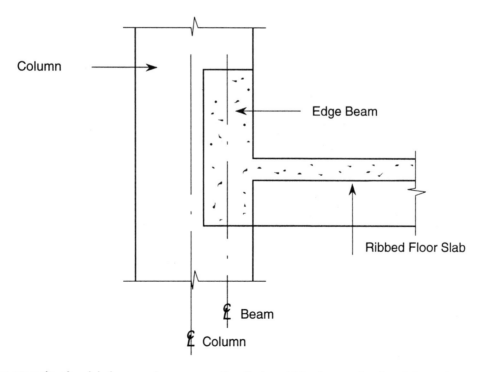

Column

Edge Beam

Ribbed Floor Slab

₵ Beam

₵ Column

Figure 4.9 An example of a slab–beam–column connection that could lead to torsional and flexural cracking.

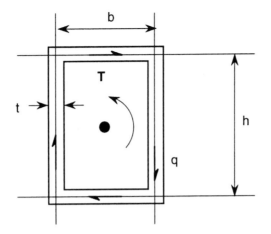

Figure 4.10 Torsional shear flow around a closed cell.

In EC2, the maximum torsional moment T_{Rd1} that can be resisted by the compressive struts in the concrete (equation (4.40), cl. 4.4.3.1, P(6)) is given by

$$T_{Rd1} = 2vf_{cd}tA_k/(\cot\theta + \tan\theta) \qquad (4.11)$$

Thus if θ is taken as $45°$ T and T_{Rd1} are identical. The notation for the thin-walled closed section given in EC2 is shown in Figure 4.11, noting that $t \leqslant A/u$ $\leqslant$ actual wall thickness. For solid sections, t denotes the equivalent thickness of the wall. The efficiency factor v is given by

$$v = 0.7(0.7 - f_{ck}/200) \not< 0.35$$
$$(f_{ck} \text{ in N/mm}^2) \qquad (4.12)$$

The value of 0.35 applies if there are stirrups only along the outer periphery of the member. If closed stirrups are provided in both sides of each wall of the equivalent hollow section or in each wall of a box section, v can be assumed to be $0.7 - f_{ck}/200$ $\not< 0.5$.

If s is the spacing of the stirrups and A_{sw} the cross-sectional area of the bars used in the stirrups, then

the stirrups may be considered to act as a thin tube of thickness $t = A_{sw}/s$.

From equation (4.8), the maximum torsional moment T_{Rd2} that can be resisted by the reinforcement is

$$T_{Rd2} = 2qA_k \qquad (\text{where } A_k = bh)$$

If f_{ywd} is the design yield strength of the stirrups,

$$f_{ywd} = q/(A_{sw}/s)$$

thus

$$T_{Rd2} = 2A_k(f_{ywd}A_{sw}/s) \qquad (4.13)$$

This equation is identical to equation (4.43) in EC2 with θ equal to $45°$.

In a similar manner, the additional area of longitudinal steel for torsion is given by

$$A_{sl}f_{yld} = T_{Rd2}u_k/2A_k \qquad (4.14)$$

where f_{yld} is the design yield strength of the longitudinal reinforcement of area A_{sl}.

EC2 gives a simplified design procedure (cl. 4.3.3.2.2.), which is outlined below.

4.5.1 Torsion combined with flexure

It should be noted that, under combined torsion and flexure, both the compressive and tensile stresses induced can be additive (Narayanan, 1986), and this should be considered in the design procedure. The procedure is as follows:

The longitudinal steel required for flexure and torsion should be determined separately in accordance with the methods given previously. In the flexural tension zone, the longitudinal torsion steel should be additional to that required to resist flexural longitudinal forces. In the flexural compression

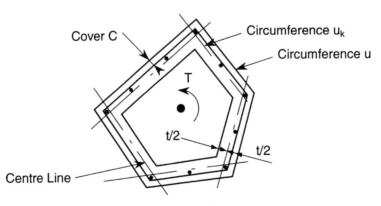

Figure 4.11 EC2 notation for torsion on the basis of a thin-walled closed section.

zone, if the tensile force due to torsion is less than the concrete compression due to flexure, no additional longitudinal torsion steel is necessary. Where torsion is combined with a large bending moment, high principal stresses can occur in the compression zone and these should not exceed αf_{cd} (the upper limit for flexure). The principal stress is obtained from the mean longitudinal compression in flexure and the tangential stress due to torsion. This is taken as

$$\tau_{sd} = T_{sd}/(2A_k t) \qquad (4.15)$$

4.5.2 Torsion combined with shear

Again, flexural and pure torsional shear stresses can be additive, and thus the following interaction formula is adopted:

$$(T_{sd}/T_{Rd1})^2 + (V_{sd}/V_{Rd2})^2 < 1 \qquad (4.16)$$

The symbols have been defined previously. EC2 requirements for flexure, shear and torsion are brought together in the following example.

Example 4.3: flexure, shear and torsion

A reinforced concrete section has a width $b_w = 400$ mm, effective depth $d = 600$ mm and overall depth $h = 650$ mm. Given that $f_{ck} = 30$ N/mm² and $f_{yk} = 460$ N/mm², (i) determine the maximum capacity in flexure, shear and torsion considered separately and (ii) estimate the reinforcement requirements for $M_{sd} = 500$ kN m, $V_{sd} = 400$ kN and $T_{sd} = 100$ kN m.

(i) The maximum flexural capacity (singly reinforced) from Figure 4.2 is

$$\begin{aligned} M_{sd} &= 0.167\, f_{ck} b_w d^2 \\ &= 0.167 \times 30 \times 400 \times 600^2 \times 10^{-6} \\ &= 721.4 \text{ kN m} \end{aligned}$$

The maximum shear capcity from equation (4.3) is

$$V_{Rd2} = 0.50\, v f_{cd} b_w \times 0.9d$$

where

$$v = 0.7 - f_{ck}/200 = 0.55$$

Thus

$$\begin{aligned} V_{Rd2} &= 0.5 \times 0.55 \times (30/1.5) \times 400 \times 0.9 \\ &\qquad \times 600 \times 10^{-3} \\ &= 1188 \text{ kN} \end{aligned}$$

The maximum torsional capacity from equation (4.10) is

$$T_{Rd1} = 2v f_{cd} t A_k$$

where

$$\begin{aligned} t &= A/u = 400 \times 650/2\,(400 + 650) \\ &= 123.8 \text{ mm} \end{aligned}$$

This is greater than the EC2 requirement of not less than twice the cover c (say $2 \times 50 = 100$ mm) to the longitudinal beams.

$$\begin{aligned} A_k &= (400 - 123.8) \times (650 - 123.8) \\ &= 0.1453 \times 10^6 \text{ mm}^2 \end{aligned}$$

$$v = 0.7\,(0.7 - f_{ck}/200) = 0.385$$

Thus

$$\begin{aligned} T_{Rd1} &= 2 \times 0.385 \times (30/1.5) \times 123.8 \times 0.1453 \\ &= 277.02 \text{ kN m} \end{aligned}$$

(ii) With $M_{sd} = 500$ kN m, then from Figure 4.3

$$M_{sd}/b_w d^2 = 500 \times 10^6/400 \times 600^2 = 3.47$$

Thus

$$x/d = 0.3$$

and

$$x = 180 \text{ mm}$$

$$d - 0.4x = 528 \text{ mm}$$

Thus

$$A_s = 500 \times 10^6/(460/1.15) \times 528 = 2367.4 \text{ mm}^2$$

Provide: 2–25φ and 2–32φ (2592 mm²).

Reinforcement ratio

$$p_1 = 2592/400 \times 600 = 0.0108$$

From equation (4.7)

$$V_{Rd1} = T_{Rd}k(1.2 + 40p_1)b_w d$$

for $f_{ck} = 30$

$$T_{Rd1} = 0.34 \text{ N/mm}^2 \qquad \text{(Table 1.6)}$$

$$k = 1.6 - 0.6 = 1.0$$

If the longitudinal bars are not curtailed, then $p_1 = 0.0108$. Thus

$$V_{Rd1} = 0.34 \, (1.2 + 40 \times 0.0108) \times 400 \times 0.9$$
$$\times 600 \times 10^3$$
$$= 119.85 \text{ kN}$$

As $V_{Rd1} = V_{cd} = 119.85$ kN and $3V_{cd}$ is less than V_{sd} (400 kN), a crack control check is required; see section 4.8. Shear reinforcement is required to resist a shear of $V_{sd} - V_{Rd1}$, thus

$$V_{wd}(\text{stirrups}) = 400 - 119.85 = 280.15 \text{ kN}$$
$$= (A_{sw}/S)0.9 d f_{ywd}$$

Adopting four legs 10 mm φ ($A_{sw} = 314$ mm^2) with $f_{ywd} = 460/1.15 = 400$ N/mm^2, then

$$s \quad \frac{= 314 \times 0.9 \times 600 \times 400}{280.15 \times 10^6}$$
$$= 242.1 \text{ mm c/c (say 240 mm c/c)}$$

The shear reinforcement ratio (see section 4.6) is

$$p_w = A_{sw}/sb_w = 314/240 \times 400$$
$$= 0.0032 > 0.0012$$

This satisfies the minimum shear reinforcement requirement (table 5.5 of EC2) and for crack control

$$(V_{sd} - 3V_{cd})/p_w b_{wd} = (400 - 3 \times 119.85)10^3/$$
$$0.0032 \times 400 \times 600$$
$$= 52.67 \text{ N/mm}^2$$

Thus from table 4.13 (EC2) the maximum stirrup spacing is $300 - (2.67/25) \times 100 = 289.32$ mm. Thus the spacing of the links for crack control is satisfactory.

The maximum torsional moment that can be resisted by the compressive struts in the concrete is 277.02 kN m and this is greater than the design torsion $T_{sd} = 100$ kN. The torsional reinforcement is obtained from equation (4.13) and with 10 mm φ links ($A_{sw} = 78.5$ mm^2) and $f_{ywd} = 400$ N/mm^2

$$s = 2A_k f_{ywd}(A_{sw}/T_{Rd2})$$
$$= 2 \times 0.1453 \times 10^6 \times 400 \times (78.5/100 \times 10^6)$$
$$= 91.2 \text{ mm (say 90 mm c/c)}$$

An arrangement of links – two legs at 90 mm c/c and six legs at 180 mm c/c (doubling up for outer leg) – will provide adequate resistance for shear and torsion.

From equation (4.14), the additional area of longitudinal steel for torsion with $f_{yld} = 400$ N/mm^2 N/mm^2 is

$$A_{sl} = (T_{Rd2}u_k/2A_k)/f_{yld}$$

where

$$u_k = 2[(400 - 123.8) + (650 - 123.8)]$$
$$= 1604.8 \text{ mm}$$

Thus

$$A_{sl} = (100 \times 10^6 \times 1604.8/2 \times 0.1453 \times 10^6)/400$$
$$= 1380.6 \text{ mm}^2 \text{ (2–20}\varphi \text{ (628) and } = 4\text{–16}\varphi \text{ (804))}$$

In the flexural zone, the longitudinal torsion steel is additional to that required to resist flexure, but in the compression zone the longitudinal torsion steel can be omitted if the tensile force due to torsion is less than the concrete compression. For reasons of simplicity, it is suggested that the longitudinal torsion steel is distributed round the inner periphery of the links with one bar at each corner with intermediate bars spaced not more than 350 mm centres (see section 4.7).

As the torsion is combined with a large bending moment, the principal stress in the compression zone should be checked. This is estimated from the mean longitudinal compression in flexure and the tangential shear stress τ_{sd} due to the torsion $T_{sd} = 100$ kN m. From equation (4.15):

$$\tau_{sd} = T_{sd}/(2A_k t)$$
$$= 100 \times 10^6/2 \times 0.1453 \times 10^6 \times 123.8$$
$$= 2.78 \text{ N/mm}^2$$

EC2 does not give any guidance on evaluating the mean longitudinal compression and it is suggested that a parabolic distribution is assumed. Thus the mean value is two-thirds the maximum value, say $0.67 \times 0.85 \times 30/1.5 = 11.39$ N/mm$^2 = f_{cm}$.

The principal compressive stress is given by

$$f_{pcs} = f_{cm}/2 \pm [(f_{cm}/2)^2 + \tau_{sd}^2]^{1/2}$$

$$= 11.39/2 \pm [(11.39/2)^2 + 2.78^2]^{1/2}$$

$$= 5.7 \pm 6.34$$

$$= 12.04 \text{ N/mm}^2$$

This value should not exceed αf_{cd}(cl. 4.3.3.2.1(1)), which is $0.85 \times 20 = 17$ N/mm².

As torsion is combined with shear, the following interaction formulae should be satisfied:

$$(T_{sd}/T_{Rd1})^2 + (V_{sd}/V_{Rd2})^2 \leqslant 1.0$$

$$(100/277.02)^2 + (400/1188)^2 = 0.13 + 0.11$$

$$= 0.24 < 1.0$$

The section on torsion in EC2 includes a note on warping torsion (this can normally be neglected).

4.6 SUMMARY OF EC2 DETAILING PROVISIONS (CL. 5.0)

4.6.1 General

Section 5 of EC2 covers detailing provisions, and space limitations necessitate the material dealt with in this text to be restricted to the more important clauses. Clause 5.2.1 refers to the spacing of bars to ensure that the concrete can be placed and compacted satisfactorily and that adequate bond can be developed together with diameters. It should be noted that table 5.1 in EC2 should be replaced by table 8 of the NAD for UK application.

4.6.2 Bond

The Code differentiates between good and poor bond conditions, the quality of bond depending on the surface pattern of the bar, the dimension of the member and the position and inclination of the reinforcement during concreting. Bond conditions are defined in Figure 4.12 and design values for the ultimate bond stress f_{bd} for good bond conditions are obtained from the following equations:

$$f_{bd} = (0.36\, f_{ck}^{1/2})/\gamma_c \quad \text{plain bars} \quad (4.17)$$

$$= (2.25\, f_{ctk,0.05})/\gamma_c \quad \text{high-bond bars} \quad (4.18)$$

These are tabulated in Table 1.6 for the nine grades of concrete ($f_{ck} = 12$–50 N/mm²) with $\gamma_c = 1.5$. For poor bond conditions, the values in Table 1.6 are multiplied by 0.7. The basic anchorage length (cl. 5.2.2.3) for reinforcement l_b is the straight length required for anchoring the force $A_s f_{yd}$, that is, for a bar of diameter φ,

$$(\pi\, \varphi^2/4)\, f_{yd} = f_{bd}\pi\, \varphi l_b$$

Thus

$$l_b = (\varphi/4)\, (f_{yd}/f_{bd}) \quad (4.19)$$

The above equation assumes a constant bond stress equal to f_{bd}.

The required anchorage length $l_{b,net}$ (see Figure 4.12) is obtained from equation (5.4) of EC2 in which l_b is modified to take account of the ratio of the reinforcement required by design to that provided, anchorage in tension and compression and whether the reinforcing bar is straight or curved.

Refer to EC2 for splices and laps (cl. 5.2.4), anchorage of links and shear reinforcement (cl. 5.2.5), additional notes for high-bond bars exceeding $\boxed{32\text{ mm}}$ in diameter (cl. 5.2.6) and bundled high-bond bars (cl. 5.2.7).

4.6.3 Beams (cl. 5.4.2)

(a) Longitudinal reinforcement (cl. 5.4.2.1)

The assessment of the minimum area of reinforcement required to ensure controlled cracking is covered in section 4.8. This value should not be less than (cl. 5.4.2.1.1)

$$\boxed{0.6}\, b_t d/f_{yk} \leqslant \boxed{0.0015}\, b_t d \quad (4.20)$$

where f_{yk} is in N/mm², b_t denotes the mean width of the tension zone – for a T beam with the flanges in compression, only the width of the web is taken into account in calculating the value of b_t. Equation (4.20) is given in tabular form in Table 4.3 for a number of grades of reinforcement and it can be seen that $0.0015 b_t d$ governs if f_{yk} exceeds 400 N/mm².

The cross-sectional area of the tension reinforcement and the compression reinforcement should not be greater than $\boxed{0.04 A_{sc}}$ other than at laps.

Other detailing requirements (see EC2) deal with the length of the longitudinal tensile reinforcement (cl. 5.4.2.1.3), anchorage of bottom reinforcement at an end support (cl. 5.4.2.1.4) and at intermediate supports (cl. 5.4.2.1.5). In monolithic construction, even when simple supports have been assumed in design, the section should be designed for bending moment arising from partial fixity of at least $\boxed{25\%}$ of the maximum bending moment in the span.

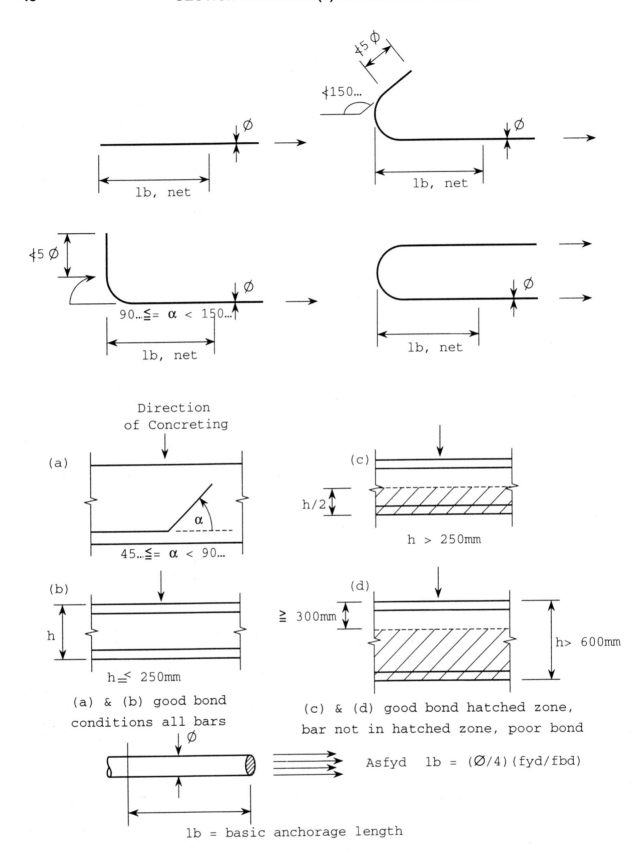

Figure 4.12 Bond requirements.

Table 4.3 Equation (4.20) in tabular form for a number of grades of reinforcement

f_{yk} (N/mm^2)	$0.6/f_{yk}$
220	0.0027
250	0.0024
400	0.0015
460	0.0013
500	0.0012

Table 4.4 Spacing of stirrups in beams for crack control if $V_{cd} > V_{sd}$ or $3V_{cd} > V_{sd}$ no check required (table 4.13 of EC2)

$(V_{sd} - 3V_{cd})/p_w b_w d$ (N/mm^2)	Stirrup spacing (mm)
<50	300
75	200
100	150
150	100
200	50

Table 4.5 Minimum values of p_w, modified to include additional steel classes (table 5.5 of EC)

Concrete classes	Steel classes				
	S220	S250	S400	S460	S500
C12/15 and C20/25	0.0016	0.0014	0.0009	0.0008	0.0007
C25/30 to C35/45	0.0024	0.0021	0.0013	0.0012	0.0011
C40/50 to C50/60	0.003	0.0026	0.0016	0.0014	0.0013

if $\quad \tfrac{1}{5}V_{Rd2} < V_{sd} \leq \tfrac{2}{3}V_{Rd2}$

$S_{max} = 0.6d \leq 300$ mm $\qquad$ (EC2 eqn. 5.18)

if $\quad V_{sd} > \tfrac{2}{3}V_{Rd2}$

$S_{max} = 0.3d \leq 200$ mm $\qquad$ (EC2 eqn. 5.19)

The symbols above have been defined previously. The transverse spacing of the legs in a link assembly should not exceed:

if $\quad V_{sd} \leq \tfrac{1}{5}V_{Rd2}$

$S_{max} = d$ or $\boxed{800\text{ mm}}$ whichever is the smaller

if $\quad V_{sd} > 1/5 \ V_{Rd2}$

equation (5.18) or (5.19) of EC2 applies

4.6.4 Torsion reinforcement

The minimum values of p_w given in Table 4.13 of EC2 also apply to torsion. The longitudinal spacing of torsion links should not exceed $U_k/8$ with the proviso that the requirements of equations (5.17) to (5.19) for maximum spacing of links are complied with. The longitudinal bars should be arranged such that there is at least one bar at each corner, the others being distributed around the inner periphery of the links, spaced at not more than $\boxed{350\text{ mm}}$ centres.

4.6.5 Cast *in situ* solid slabs (cl. 5.4.3)

Minimum bar spacings and maximum bar diameters to control cracking without direct calculation are covered in section 4.8. In addition, the following rules apply to two-way and one-way solid slabs. The absolute minimum thickness is $\boxed{50\text{ mm}}$ and, in the main direction, the minimum and maximum steel percentages of flexural reinforcement are as for

(b) Shear reinforcement

The following summarizes the detailing rules for shear reinforcement consisting of links enclosing the longitudinal tensile reinforcement and the compression zone. For other forms of shear reinforcement, refer to EC2 (cl. 5.4.2.2)

The shear reinforcement ratio is defined as:

$p_w = A_{sw}/(s b_w \sin \alpha)$

s and b_w have been defined previously and α is the angle between the shear reinforcement and the main steel. Sin $\alpha = 1.0$ for vertical links. To control cracking due to tangential action effects, the stirrup spacing is related to the shear capacity in the concrete compression zone V_{cd} as given in Table 4.13 of EC2; see Table 4.4. The value of V_{cd} may be taken as V_{Rd1}.

In section 5 of EC2, minimum values of P_w are related to the concrete class and steel class. The steel classes given in EC2 are S220, S400 and S5500 and in Table 4.5, two additional classes have been added.

The diameter of shear reinforcement should not exceed $\boxed{12\text{ mm}}$ where it consists of plain round bars. The maximum longitudinal spacing (S_{max}) of successive series of stirrups is governed by the following expressions:

if $\quad V_{sd} \leq \tfrac{1}{5}V_{Rd2}$

$S_{max} = 0.8d \leq 300$ mm $\qquad$ (EC2 eqn. 5.17)

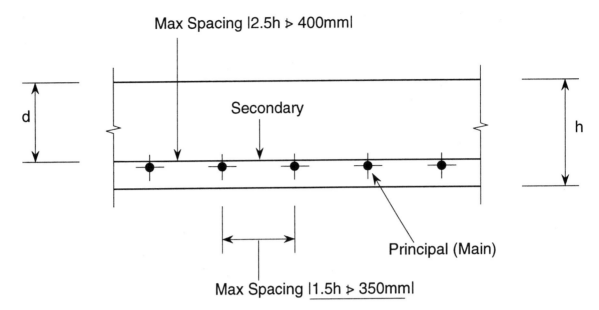

Figure 4.13 Bar spacing in slabs.

beams. Secondary (transverse) reinforcement should be provided to a value of at least 20% of the principal reinforcement. The maximum spacing of the bars is shown in Figure 4.13. Refer to EC2 for reinforcement in slabs near supports (cl. 5.4.3.2.2), corner reinforcement (cl. 5.4.3.2.3), reinforcement at free edges (cl. 5.4.3.2.4) and shear reinforcement (cl. 5.4.3.3).

4.7 BEAMS – DESIGN FOR SEISMIC ACTIONS

The analysis of structures for seismic actions is covered in Chapter 7, and in this section, the design of beams for given seismic actions is considered. Briefly, the seismic action effects incorporating the dead load (G), the imposed load (Q) multiplied by a combination factor ψ and the seismic loading E $(G + E + \psi Q)$ may be evaluated for regular frameworks (see Chapter 7) using a simplified dynamic analysis. The seismic action E is related to a seismic coefficient, which is dependent on the product of the amplification factor, the peak ground acceleration, the soil factor and the period *divided* by a behaviour factor q. The parameter q is introduced to account for the energy dissipation capacity and post-elastic resistance of the structure. Broadly, the behaviour factor q increases with increase in ductility of the structure. EC8 considers three ductility classes – low (L), medium (M) and high (H). The behaviour factor q is related to q_0, a basic value of the behaviour factor, dependent on the

structural type (e.g. frame, wall system, etc.), a factor k_D reflecting the ductility class (L, M, H), a factor k_R reflecting the structural regularity and a factor k_W reflecting the prevailing failure mode in structural systems with walls. For a regular structure with a framed system, k_R and k_W are taken as unity, $q_0 = 5.0$ and thus q can be expressed as $q = 5.0k_D$ where k_D is 1.0, 0.75 and 0.5 for ductility levels H, M and L respectively. Thus values of q for a regular framework are as below:

DC	q
H	5.0
M	3.75
L	2.5

Clearly, the design seismic action decreases with increase in ductility, but with increase in ductility there is a corresponding increase in complexity of design and detailing requirements. For the three levels of ductility, the detailing requirements are summarized in Table 4.6 and Figures 4.14 and 4.15, and both design and detailing requirements are compared in the following example.

Example 4.4: beam design, ductility levels

The bending moment diagram in Figure 4.16 has been obtained from a seismic analysis of a framed structure for ductility classes L, M and H. Using the

Table 4.6 EC8 summary of detailing requirements; see also Figures 4.14 and 4.15

Design action effects	Design resistance evaluation and strength	Detailing (notation as in Figure 4.14)	DC
Analysis to EC8 for seismic load combinations, capacity design criterion	Flexural strength to EC2 $V_{cd} = 0$ in critical regions, elsewhere as EC2	$l_{cr} = 2.0h_w$, $d_{bw} \geqslant 6$ mm $S_w = \min (h_w/4; 24d_{bw}; 150$ mm; $5d_{bL})$ $d_{bL} \leqslant 4.0(f_{ctm}/f_{yd})(1 + 0.8v_d)b_c$ for internal beam/column joints; for external, replace 4.0 by 5.5	H
Analysis to EC8 for seismic load combinations, capacity design criterion	Flexural strength to EC2 $V_{cd} = 40\%$ of EC2 value in critical regions, elsewhere 100%	$l_{cr} = 1.5h_w$, $d_{bw} \geqslant 6$ mm $S_w = \min (h_w/4; 24d_{bw}; 200$ mm; $7d_{bL})$ $d_{bL} \leqslant 4.5(f_{ctm}/f_{yd})(1 + 0.8v_d b_c$ for internal beam/column joints; for external, replace 4.5 by 6.0	M
Analysis to EC8 for seismic load combinations	Flexure and shear to EC2	$l_{cr} = 1.0h_w$	L

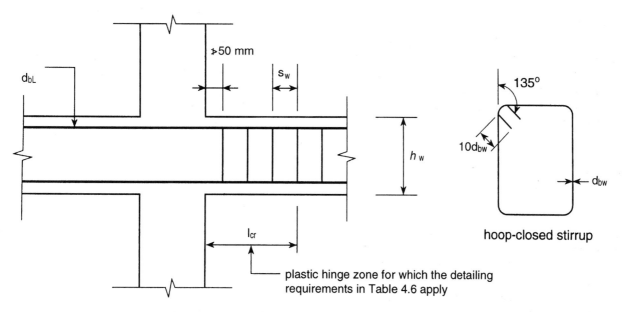

Figure 4.14 EC8 detailing requirements; see also Table 4.6 and Figure 4.15.

data provided, design the section at support B for ductility levels L, M and H. As indicated in Chapter 1, the characteristics of the steel reinforcement required in EC2 differ from those in EC8. The requirements for reinforcing steel in EC8 are summarized in Table 4.7, their aim being to ensure high ductility, high resistances after cover spalling and reliable control of designed inelastic mechanisms through capacity design procedures.

Except for closed stirrup or cross-ties, only deformed (high-bond) bars are allowed at critical sections. In this example, a characteristic strength of $f_{yk} = 400$ N/mm² is adopted and it is assumed that the characteristics given in Table 4.7 are complied with. f_{ck} is taken as 35 N/mm², noting that in EC8 the use of concrete class lower than C16 for DC (L) or C20 for DC (M) and DC (H) is not allowed.

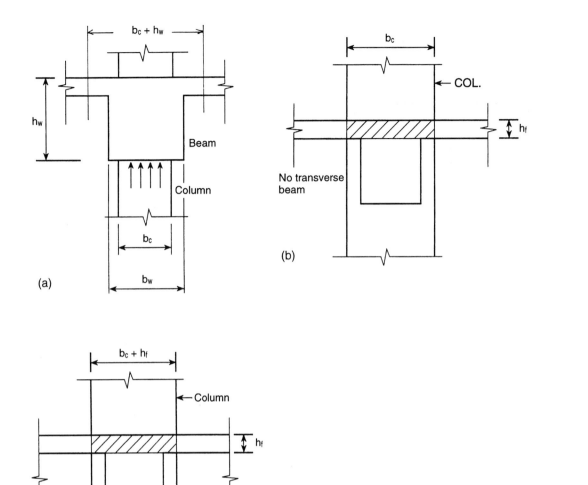

Figure 4.15   EC8 (Draft) detailing requirements for beams; see also Table 4.6 and Figure 4.14.

Ductility level 'L'

DC (L) corresponds to structures designed and dimensioned in accordance with EC2, supplemented by rules enhancing available ductility.

For DC (L), the negative and positive bending moments at support B are as follows (see Figure 4.16):

$$M_{B(L)} = 525 + 206 = 731 \text{ kN m (negative)}$$

$$M_{B(L)} = 525 - 206 = 319 \text{ kN m (positive)}$$

The analysis of the section follows the procedure set out in EC2. Thus for flexure, use can be made of the design chart; Figure 4.3. Assuming no redistribution of moment (although redistribution of moment is permitted in EC8), then for an effective depth of, say, $700 - 60 = 640$ mm and a negative bending moment of 731 kN m,

$$M_{B(L)}/b_w d^2 = 731 \times 10^6/400 \times 640^2$$
$$= 4.46$$

From Figure 4.3 with $f_{ck} = 35$ N/mm² we have

$$x/d = 0.33$$
$$x = 211.2$$
$$d - 0.4x = 555.52 \text{ mm}$$

Thus

$$A_s = 731 \times 10^6/(400/1.15) \times 555.52$$
$$= 3783 \text{ mm}^2$$

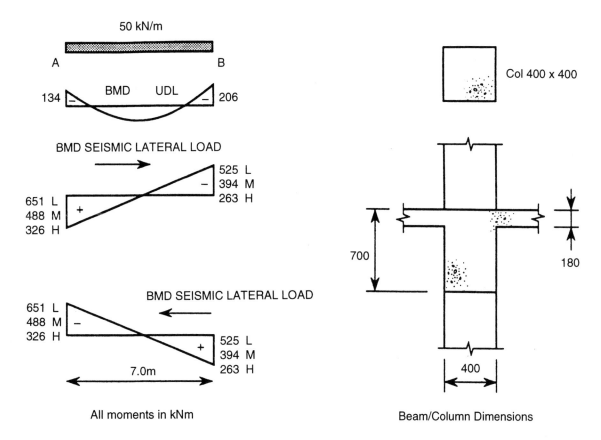

Figure 4.16 Example 4.4: bending moment diagram from seismic analysis.

Table 4.7 EC8 summary of requirements for reinforcing steel in critical regions

Properties	DC (L)	DC (M)	DC (H)
Characteristic value of uniform elongation at maximum load (%)		⩾ 6.0	⩾ 9.0
Tensile strength f_t to yield strength f_y ratio (f_t/f_y)	high-ductility steel as EC2	⩾ 1.15	⩾ 1.2
Mean values		⩽ 1.3	⩽ 1.2
Actual $f_{y, act}$ to nominal yield strength $f_{y,nom}$ ratio $(f_{y, act}/f_{y,nom})$			
Mean values		⩽ 1.25	⩽ 1.2

In order to prevent bond failure, EC2 puts restrictions on the diameter d_{bL} of longitudinal bars of beams anchored along the beam – column joint (see Table 4.6), which are related to the ductility level being considered. Assuming joint B is internal, then

$$d_{bL} \leqslant 5.0(f_{ctm}/f_{yd})(1 + 0.8\, v_d)b_c$$

where $f_{yd} = 400/1.15 = 347.8$ N/mm², $f_{ctm} = 3.2$ N/mm² from Table 1.6, $v_d = N_{sd}/f_{cd}A_c$ normalized minimal axial force in the column for seismic load combinations and b_c = width of column parallel to bars = 400 mm. In this case,

$$d_{bL} = 18.4(1 + 0.8v_d)$$

Thus if v_d is of small order, then the maximum value of d_{bL} is, say, 18 mm. Clearly, a high axial load in the column will assist in the prevention of bond failure and d_{bL} can be increased. To use 25 mm diameter bars (8 No. = 3930 mm² for A_s = 3783 mm²), the value of v_d would have to be not less than 0.5.

The shear force at B is

$$V_{B(L)} = 50 \times 3.5 + (525 + 651)/7$$
$$+ (206 - 134)/7$$
$$= 353.29 \text{ kN} = V_{sd}$$

Using $A_s = 3930$ mm², the steel ratio is

$$p = A_s/b_w d = 3930/400 \times 640$$
$$= 0.0154$$

For DC (L), EC8 puts an upper limit on the tension reinforcement ratio (p_{max}) of 75% of the maximum reinforcement ratio allowed in EC2. This applies in the critical regions of the member. From Table 4.1 with f_{ck} = 35 N/mm² and f_{yk} = 400 N/mm², $p_{max(EC2)}$ = 0.02. Thus $0.75p_{max(EC2)}$ = 0.0158. This requirement is just met with A_s = 3930 mm².

For shear resistance evaluation and verification, EC2 applies, and, assuming that the longitudinal force in the beams is of small order, then the design shear resistance of the concrete (equation (4.2)) is

$$V_{Rd1}(V_{cd}) = [\tau_{Rd}k\,(1.2 + 40p_1)]b_w d$$

where for f_{ck} = 35 N/mm², τ_{Rd} = 0.37 (see Table 1.6), k = 1.0 and $p_1 \leqslant 0.02$

$$V_{Rd1} = [0.37\,(1.2 + 40 \times 0.0154)]\,400 \times 640$$
$$\times 10^{-3}$$
$$= 172.01 \text{ kN} < V_{sd}$$

From equation (4.3)

$$V_{Rd2} = \tfrac{1}{2}vf_{cd}b_w \times 0.9d$$

where

$$v = 0.7 - f_{ck}/200 = 0.525 \geqslant 0.5$$

Thus

$$V_{Rd2} = 0.5 \times 0.525 \times (35/1.5) \times 400 \times 0.9$$
$$\times 640 \times 10^{-3}$$
$$= 1411.2 \text{ kN} > V_{sd}$$

Thus shear reinforcement is required to the value

$$V_{wd} = 353.29 - 172.01 = 181.28 \text{ kN}$$

From equation (4.5) the spacing of 8 mm φ stirrups (four legs) is

$$s_h = A_{sw} \times 0.9df_{ywd}/_{wd}$$
$$= 201 \times 0.9 \times 640 \times (400/1.15)/181.28 \times 10^3$$
$$= 222.1 \text{ mm (say 220 mm c/c)}$$

The shear reinforcement ratio, see section 4.6.3(b), is

$$p_w = A_w/s_w b_w = 201/220 \times 400 = 0.0023$$

This meets the requirements of Table 5.5 of EC2 for minimum values of p_w. As $3V_{Rd1}$ = 544.21 kN is greater than V_{sd} = 353.29 kN, no check is required for crack control; see Table 4.13 of EC2.

A final requirement of EC8 (applies to DC (L), (M) and (H)) is that the tension steel ratio should nowhere be less than

$$p_{min} = 0.5(f_{ctm}/f_{yk})$$

In this case p_{min} = 0.5 × 3.2/400 = 0.004

Thus, it can be seen that section analysis for flexure and shear to DC (L) follows the procedure given in EC2 with some additional requirements to ensure reasonable ductility at critical regions. General detailing requirements for the three ductility levels are given in Table 4.8.

Ductility level 'M'

DC (M) corresponds to structures designed, dimensioned and detailed according to specific earthquake-resistant provisions, enabling the structure to enter well within the inelastic range under reputed reversed loading, without suffering brittle failure.

Table 4.8 Summary of EC8 requirements for minimum (p_{min}) and maximum (p_{max}) reinforcement ratios[a]

f_{ck} (N/mm²)	20	25	30	35	40	45	50
f_{ctm} (N/mm²) from EC2	2.2	2.6	2.9	3.2	3.5	3.8	4.1
p_{min} (f_{yk} = 250)	0.0044	0.0052	0.0058	0.0064	0.0070	0.0076	0.0082
p_{min} (f_{yk} = 400)	0.0028	0.0033	0.0036	0.0040	0.0044	0.0048	0.0051
p_{min} (f_{yk} = 460)	0.0024	0.0028	0.0031	0.0035	0.0038	0.0041	0.0044
p_{min} (f_{yk} = 500)	0.0022	0.0026	0.0029	0.0032	0.0035	0.0038	0.0041

p_{max} for DC (L): within critical regions ⩽75% EC2 values (see Table 4.1 for EC2 values)
p_{max} for DC (M): $0.65(f_{cd}/f_{yd})p^1/p + 0.0015$ within critical regions (p^1 is the compression steel ratio)
p_{max} for DC (H): $0.35(f_{cd}/f_{yd})p^1/p + 0.0015$ within critical regions
For DC (M) and DC (H) at least two 14 mm φ bars (f_{yk} = 400 N/mm²) to be provided on the top and bottom faces along the entire length of the beam

[a]Values of p_{min} related to f_{ctm} and f_{yk} are the same for three levels of ductility.

From Figure 4.16, the negative and positive moments for DC (M) at support B are 600 kN m and 188 kN m respectively. From Figure 4.3

$$M_{B(M)}/b_w d^2 = 600 \times 10^6/(400 \times 640^2) = 3.66$$

Thus

$$x/d = 0.27$$

$$x = 172.8$$

$$d - 0.4\,x = 570.9$$

$$A_s = 600 \times 10^6/[(400/1.15) \times 570.9] = 3021 \text{ mm}^2$$

For DC (M), the value of d_{bL} for an internal beam is

$$d_{bL} \leqslant 4.5\,(f_{ctm}/f_{yd})(1 + 0.8\,v_d)b_c$$

and the notation is as for DC (L).

Using 10–20φ bars ($A_s = 3140$ mm^2), the steel ratio is

$$p = A_s/b_w d = 3140/400 \times 640 = 0.0123$$

Within the critical regions, the tension reinforcement ratio p_{max} should not exceed

$$p_{max} = 0.65(f_{cd}/f_{yd})p^1/p \times 0.0015$$

where p^1 is the compression steel ratio of the beam bars passing through the joint.

In this case, $f_{cd} = 35 \times 0.85/1.5 = 19.83$ N/mm^2 and $f_{yd} = 400/1.15 = 347.83$ N/mm^2. Thus $f_{cd}/f_{yd} = 0.057$ and values of p_{max} for $p^1/p = 0.25, 0.5, 0.75$ and 1.0 are given below:

p_{max}	p^1/p
0.011	0.25
0.20	0.5
0.29	0.75
0.039	1.0

Thus the influence of the presence of compression reinforcement in the critical regions on ductility is recognized. A further requirement is that at least two 14 mm diameter bars ($f_{yk} = 400$ N/mm^2) should be provided both at the top and bottom faces along the entire length of the beam.

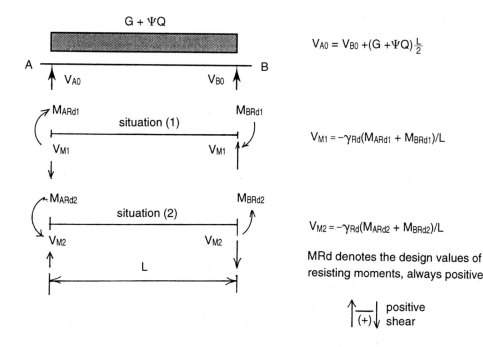

Figure 4.17 EC2 (Draft) evaluation of maximum and minimum design shear forces.

For DC (M) (and DC (H)), EC8 requires that the design shear forces are to be determined in accordance with *capacity design criterion*. This means that moments are derived from the actual areas of tension steel and a factor γ_{Rd} is introduced that is intended to counterbalance the partial safety factor γ_s for steel and cover strain hardening effects. In the absence of more precise data, γ_{Rd} may be taken as 1.25.

In this example, $A_{s(provided)}/A_{s(required)}$ at support B is 3140/3021 = 1.04, and to allow for this, the shear due to hyperstatic effects will be calculated with, say, a 5% increase in the moments shown in Figure 4.16. Evaluation of the capacity design values of shear forces is obtained from Figure 4.17. Thus

$$V_{A0} = V_{B0} = 50 \times 3.5 = 175 \text{ kN}$$

$$M_{ARd1} = 488 \times 1.05 = 512.4 \text{ kN m}$$

$$M_{BRd1} = 394 \times 1.05 = 413.7 \text{ kN m}$$

$$M_{ARd2} = 488 \times 1.05 = 512.4 \text{ kN m}$$

$$M_{BRd2} = 394 \times 1.05 = 413.7 \text{ kN m}$$

$$V_{M1} = -1.25 \ (512.4 + 413.7)/7 = -165.4 \text{ kN}$$

$$V_{M2} = -1.25 \ (512.4 + 413.7)/7 = 165.4 \text{ kN}$$

Thus at B, the maximum shear force $V_{s,max} = 175 + 165.4 = 340.4$ kN and the minimum shear force $V_{s,min} = 175 - 165.4 = 9.6$ kN. If $\xi = V_{s,min}/V_{s,max}$ is greater than or equal to –0.5, then the shear resistance is computed as in EC2, which applies in this case as $\xi = +0.028$, which is greater than –0.5. For ξ less than –0.5, reference should be made to EC8, and where additional detailing requirements are necessary to control sliding shear failure or excessive bidiagonal cracking. For critical regions, V_{Rd1} (V_{cd}) is taken as 40% of the EC2 value, and elsewhere 100%.

From equation (4.2)

$$0.4 V_{Rd1}(V_{cd}) = 0.4 \ [\tau_{Rd} k (1.2 + 40 \ p_1)] b_w d$$
$$= 0.4 \ [0.37 \ (1.2 + 40 \times 0.0123)]$$
$$400 \times 640 \times 10^{-3}$$
$$= 64.1 \text{ kN}$$

and

$$V_{Rd2} = 1411.2 \text{ kN} > V_{sd} \ (340.4 \text{ kN})$$

Thus shear reinforcement is required to the value

$$V_{wd} = 340.4 - 64.1 = 276.3 \text{ kN}$$

From equation (4.5), the spacing of 8 mm φ stirrups (four legs) is

$$s_w = A_{sw} \times 0.9 \ df_{ywd}/V_{wd}$$
$$= 201 \times 0.9 \times 640 \times (400/1.15)/276.3 \times 10^{-3}$$
$$= 145.7 \text{ (say 140 mm c/c)}$$

Referring to Table 4.6, the diameter of the stirrups d_{bw} should not be less than 6 mm and s_w should not exceed the minimum of

$$h_w/4 \quad = 700/4 = 175 \text{ mm}$$

$$24 \ d_{bw} = 24 \times 8 = 192 \text{ mm}$$

$$7 \ d_{bL} \quad = 7 \times 20 = 140 \text{ mm}$$
$$\text{or} \qquad 200 \text{ mm}$$

Thus $7d_{bL}$ (140 mm) governs.

Ductility level 'H'

DC (H) corresponds to structures for which the design, dimensioning and detailing provisions are such as to ensure, in response to the seismic excitation, the development of chosen stable mechanisms associated with large hysteretic energy dissipation. From Figure 4.16, the negative and positive moments at support B are 469 kN m and 57 kN m respectively. From Figure 4.3

$$M_{B(H)}/b_w d^2 = 469 \times 10^6/400 \times 640^2 = 2.86$$

Thus

$$x/d = 0.24$$

$$x = 153.6 \text{ mm}$$

$$d - 0.4 \ x = 578.6 \text{ mm}$$

$$A_s = 469 \times 10^6/(400/1.15) \times 578.6 = 2330 \text{ mm}^2$$

For DC (H), the value of d_{bL} for an interior beam/column joint is

$$d_{bL} \leq 4.0(f_{ctm}/f_{yd}) \ (1 + 0.8\nu_d) \ b_c$$

Using 5–20φ and 3–16φ ($A_s = 2353$ mm²), the steel ratio is

$$p = A_s/b_w d = 2353/400 \times 640 = 0.0092$$

Within the critical regions, the tension reinforcement ratio p_{max} should not exceed

$$p_{max} = 0.35 \, (f_{cd}/f_{yd})p^1/p + 0.0015$$

As for DC (M), the value of f_{cd}/f_{yd} is 0.057 and values of p_{max} for $p^1/p = 0.25, 0.5, 0.75$ and 1.0 are given below:

P_{max}	p^1/p
0.0065	0.25
0.0011	0.5
0.016	0.75
0.021	1.0

Clearly, the greater ductility requirements for DC (H) necessitate the use of lower values of p_{max} than for DC (M). A further requirement is that at least two 14 mm diameter bars should be provided both at the top and bottom faces along the entire length of the beam.

As stated previously, EC8 requires that the design shear forces are to be determined in accordance with the capacity design criterion. For DC(H) $A_{s(provided)}/A_{s(required)}$ is 2353/2330 = 1.01 and thus there is negligible error in using the moments given in Figure 4.16 for DC (H). From Figure 4.17

$$V_{A0} = V_{B0} = 175 \text{ kN}$$

$$M_{ARD1} = 326 \text{ kN m}$$

$$M_{BRd1} = 263 \text{ kN m}$$

$$M_{ARd2} = 326 \text{ kN m}$$

$$M_{BRd2} = 263 \text{ kN m}$$

Taking γ_{Rd} as 1.25,

$$V_{M1} = -1.25 \, (326 + 263)/7 = -105.2 \text{ kN}$$

$$V_{M2} = 1.25 \, (326 + 263)/7 = 105.2 \text{ kN}$$

Thus at B, the maximum shear force $V_{s,max} = 175 + 105.2 = 280.2$ kN and the minimum value of the shear force $V_{s,min} = 175 - 105.2 = 69.8$ kN. As $V_{s,min}/V_{s,max}$ is greater than -0.5, then the shear resistance is calculated as in EC2, but in the critical regions, $V_{Rd1}(V_{cd})$ is taken as zero, and elsewhere as in EC2. Thus shear reinforcement is required to the value of

$$V_{wd} = 280.2 \text{ kN}$$

From equation (4.5), the spacing of 8 mm φ stirrups (four legs) is

$$\begin{aligned} s_w &= A_{sw} \times 0.9 \, df_{ywd}/V_{wd} \\ &= 201 \times 0.9 \times 640 \times (400/1.15)/280.2 \times 10^3 \\ &= 143.7 \text{ mm (say 140 mm)} \end{aligned}$$

Referring to Table 4.6, the diameter of the stirrups d_{bw} should not be less than 6 mm and s_w should not exceed the minimum of

$$h_w/4 = 700/4 = 175 \text{ mm}$$

$$24 \, d_{bw} = 24 \times 8 = 192 \text{ mm}$$

$$5 \, d_{bL} = 5 \times 16 = 80 \text{ mm}$$
or 150 mm

Thus $5d_{bL}$ governs (80 mm) with d_{bL} equal to 16 mm φ.

Summary

The reinforcement requirements at end B for the three levels of ductility are given in Table 4.9. Clearly, a similar calculation is required at end A but is omitted for brevity.

It can be seen from Table 4.6 that there is a significant reduction in the required longitudinal steel ratio with increase in ductility of the section. This requires a corresponding increase in the amount of confinement reinforcement and thus a closer spacing of the stirrups. Reference should also be made to Figures 4.14 and 4.15 and Tables 4.6 and 4.8 for general detailing requirements. In order to take advantage of the beneficial effect of the column

Table 4.9 Summary of beam reinforcement requirements at end cross-section B (Example 4.4)

Ductility class	Longitudinal tension reinforcement	Shear reinforcement (vertical stirrups)
Low (L) $q = 2.5$	3930 mm² $p = 0.0154$	8 mm φ (four legs) $s_w = 220$ mm $p_w = 0.0023$
Medium (M) $q = 3.75$	3140 mm² $p = 0.0123$	8 mm φ (four legs) $s_w = 140$ mm $p_w = 0.0036$
High (H) $q = 5.0$	2353 mm² $p = 0.0092$	8 mm φ (four legs) $s_w = 80$ mm $p_w = 0.0063$

compressive force on the bond of horizontal bars through the joint, the beam width is limited to $b_w \leq b_c + h_w$ (see Figure 4.15(a)), and in no case should b_w exceed $2b_c$. EC8 recommends that top reinforcement at the end cross-sections of T or L beams should be placed mainly within the beam width. Some of the reinforcement may be placed within the hatched zones shown in Figures 4.15(b) and (c). The effective flange width b_e is as follows:

$$b_e = b_c$$

exterior columns, no transverse beams

$$= b_c + 2\,h_f$$

exterior columns with transverse beams

$$= b_c + 2\,h_f$$

interior columns, no transverse beams

$$= b_c + 4\,h_f$$

interior columns with transverse beams

It is obviously preferable to limit eccentricity between beam and column axes to a minimum, and EC8 gives an upper limit of $e \leq b_c/4$ (see Figure 4.15(d)) to ensure efficient transfer of cyclic moments from beam to column.

Reference should be made to EC8 for further provisions for anchorage in exterior beam–column joints and anchorage length of column bars. Chapter 5 deals with joint design and Chapter 6 with confinement reinforcement in columns.

Again, the three levels of ductility are considered and, as with beam end cross-sections, the amount of confinement reinforcement increases with increase in ductility.

4.8 SERVICEABILITY LIMIT STATES (CL. 4.4)

4.8.1 Stress limitations (cl. 4.4.1)

As with BS 8110, EC2 covers crack control and deflection with and without direct calculation. In addition, EC2 imposes stress limitations (cl. 4.4.1), which are summarized below.

1. It may be appropriate to limit the compressive stress to $\boxed{0.6}\,f_{ck}$ under the rare combination of loads in areas exposed to environments in class 3 or 4 (see Chapter 2 of this text) in order to prevent longitudinal cracks occurring.
2. Creep may exceed the amounts predicted using the methods given in clause 2.5.5 of EC2 if the

stress in the concrete under *quasi-permanent* loads exceeds $\boxed{0.45}\,f_{ck}$. If creep is likely to affect significantly the functioning of the member considered, the stresses should be limited to this value. For reinforced concrete flexural members, this check should be considered if the span/effective depth ratio exceeds 85% of the value given in table 4.13 of EC2.
3. A limitation of $\boxed{0.81}\,f_{yk}$ is put on the steel stress under serviceability conditions in order to avoid the development of permanently open cracks. The steel stress should be estimated using the *rare* combination of loads (see Chapter 1 of this text).
4. The stress limitations in 1, 2 and 3 above may generally be assumed to be satisfied without further calculations provided:

    *   The design for the ULS has been carried out in accordance with the Code requirements (section 4.3);
    *   the minimum reinforcement requirements are satisfied;
    *   redistribution $\not> 30\%$ at ULS.

5. Long-term effects may be taken into account by assuming a modular ratio of 15 for situations where more than 50% of the stress arises from *quasi-permanent* actions. Otherwise, they may be ignored.
6. Stresses are checked using section properties corresponding to either the uncracked or the fully cracked condition, whichever is appropriate.
7. If the maximum tensile stress in the concrete calculated on the basis of an uncracked section under the *rare* combination of loads exceeds f_{ctm}, the cracked state should be assumed.
8. Where a cracked section is used, and concrete is assumed to be elastic in compression, but incapable of sustaining tension, no allowance should be made for the stiffening effect of the concrete in tension after cracking when checking stresses.

4.8.2 Stress limitations – calculation procedure

Calculations for estimating steel and concrete stresses under serviceability conditions can be based on equations (4.21) to (4.26) for the elastic analysis of reinforced concrete sections given in Figure 4.18). For convenience, the relationship between the steel percentage p ($p = 100A_s/b_wd$), the neutral axis factor n ($n = x/d$) and the modular ratio α_e ($\alpha_e = E_s/E_c$) is reproduced in tabular form in Figure 4.19. The procedure is quantified in the following example.

Reference should also be made to Examples 4.6 and 4.7.

Example 4.5: steel and concrete stresses

The midspan section of a simply supported reinforced concrete slab of span $L = 6.0$ m designed for the ultimate limit state ($b_w = 1000$ mm, $h = 230$ mm, $d = 200$ mm, $f_{ck} = 30$ N/mm^2, $f_{yk} = 460$ N/mm^2, $G_k = 5.5$ kN/m, $Q_k = 5$ kN/m, $\gamma_{fG} = 1.35$ and $\gamma_{fQ} = 1.5$) has a steel percentage ($p = A_s/b_w d$) of 0.45. Determine the steel and concrete stresses under serviceability conditions and check for compliance with stress limitations. The values of ψ for the rare and quasi-permanent combination of loads may be taken as 0.7 and 0.3 respectively (offices and stores, table 1 of NAD).

Considering the *rare* combination of loads, the midspan bending moment is

$$M_{(SLS)} = (G_k + \psi_0 Q_k) L^2/8$$

$$= (5.5 + 0.7 \times 5.0) \times 6^2/8$$

$$= 40.5 \text{ kN m}$$

From Table 1.6, the value of $E_{cm} = 32$ kN/mm^2 for $f_{ck} = 30$ N/mm^2 and $E_s = 200$ kN/mm^2. Thus the modular ratio $\alpha_e = E_s/E_{cm} = 200/32 = 6.25$. Note that as the *quasi-permanent* action ($\psi_{122} G_k = 0.3 \times 5 = 1.5$ kN/m) will induce less than 50% of the stress under serviceability conditions, long-term effects can be ignored. The value of f_{ctm} from Table 1.6 is 2.9 N/mm^2 and an approximation to the tensile stress under serviceability conditions may be obtained from the gross section properties of the concrete (ignoring reinforcement). Thus tensile stress in concrete σ_t is

$$\sigma_t = 6 M_{(SLS)}/b_w h^2$$

$$= 6 \times 40.5 \times 10^6/10^3 \times 230^2$$

$$= 4.59 \text{ N/mm}^2 \quad \text{(thus section cracked)}$$

From equation (4.23), the value of n is obtained from

$$p = (50/\alpha_e)n^2/(1-n)$$

$$0.45 = (50/6.25)n^2/(1-n)$$

$$0 = n^2 + 0.056\,25n - 0.056\,25$$

Solving the quadratic gives

$$n = -0.056\,25/2 \pm 1/2\,(0.00316 + 0.225)^{1/2}$$

$$= 0.21$$

Referring to the graphs in Figure 4.19, n approximates to 0.21. Thus

$$1 - n/3 = 0.93$$

The steel stress σ_s from equation (4.25) with $A_s = 0.0045 \times 10^3 \times 200 = 900$ mm^2 is

$$\sigma_s = M_{(elastic)}/A_s d(1-n/3)$$

$$= 40.25 \times 10^6/900 \times 200 \times 0.93$$

$$= 242 \text{ N/mm}^2 \quad (< 0.8 f_{yk})$$

The concrete stress σ_c from equation (4.26) is

$$\sigma_c = 2M_{(elastic)}/b_w d^2 n\,(1-n/3)$$

$$= 2 \times 40.25 \times 10^6/10^3 \times 200^2 \times 0.21 \times 0.93$$

$$= 10.4 \text{ N/mm}^2 \quad (< 0.6f_{ck})$$

If the value of α_e is increased to 15, the neutral axis factor n increases to about 0.375 and thus $1-n/3 = 0.887$, giving $\sigma_s = 254$ N/mm^2 and $\sigma_c = 6.71$ N/mm^2.

These are acceptable values, well below the EC2 stress limits under serviceability conditions. However, if, as in the case of a plastic analysis of a continuous slab for ultimate limit state, the ratio of plastic moment to serviceability moment is high, there will be a significant increase in σ_c and σ_s; see Examples 4.6 and 4.7.

4.8.3 Limit states of cracking (cl. 4.4.2)

(a) General considerations (cl. 4.4.2.1)

It is a requirement of EC2 that cracking should be limited to a level that will not impair the proper functioning of the structure or cause its appearance to be unacceptable. Cracking is almost inevitable in reinforced concrete structures subject to bending, shear, torsion and tension resulting from either direct loading or restraint of imposed deformations. Cracks may also arise from other causes such as plastic shrinkage or expansive chemical reactions.

In the absence of specific requirements, it may be assumed that, for exposure classes 2–4 (see Chapter 2), limitation of the maximum design crack width to about 0.3 mm under the quasi-permanent combination of loads will generally be satisfactory for reinforced concrete members in buildings with respect to appearance and durability. For other exposure classes, refer to EC2, clause 4.4.2.1.

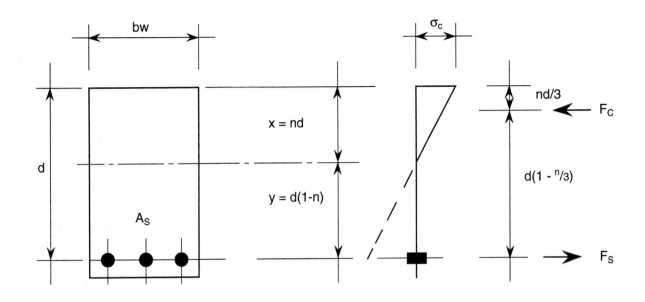

$$F_C = \frac{\sigma_c}{2} \, b_w \, nd \qquad\dots\dots\dots\dots\dots\dots\dots\dots\dots\dots\dots\dots\dots\dots \qquad (4.21)$$

$$F_S = A_S \, \sigma_c \qquad\dots\dots\dots\dots\dots\dots\dots\dots\dots\dots\dots\dots\dots\dots\dots\dots\dots \qquad (4.22)$$

$$P = \frac{100 A_S}{b_w d} = \frac{50}{\alpha e} \frac{n^2}{1-n} \qquad\dots\dots\dots\dots\dots\dots\dots\dots\dots \qquad (4.23)$$

$$\alpha e = \frac{E_S}{E_C} \qquad\dots\dots\dots\dots\dots\dots\dots\dots\dots\dots\dots\dots\dots\dots\dots\dots \qquad (4.24)$$

$$\sigma_s = \frac{M_{(elastic)}}{A_S d(1 - {}^n/3)} \qquad\dots\dots\dots\dots\dots\dots\dots\dots\dots\dots\dots \qquad (4.25)$$

$$\sigma_c = \frac{2 M_{(elastic)}}{b_w d^2 n(1 - {}^n/3)} \qquad\dots\dots\dots\dots\dots\dots\dots\dots\dots\dots \qquad (4.26)$$

Figure 4.18 Equations for elastic analysis of reinforced concrete sections; see also Figure 4.19.

(b) Minimum reinforcement areas (cl. 4.4.2.2)

EC2 distinguishes between two possible mechanisms by which tensile stresses can occur that may lead to cracking. These are:

- restraint of intrinsic (internal) imposed deformations, for example, stresses induced by restraint to shrinkage movement; and
- restraint of extrinsic (external) imposed deformations, for example, settlement of a support.

In addition, EC2 distinguishes between two basic types of stress distribution:

- bending, where the tensile stress distribution within a section is triangular, that is, part of the section remains in compression; and
- tension, where the whole of the section is subjected to tension.

In the absence of rigorous calculations, the required minimum areas of reinforcement may be calculated from equation (4.78) of EC2, that is

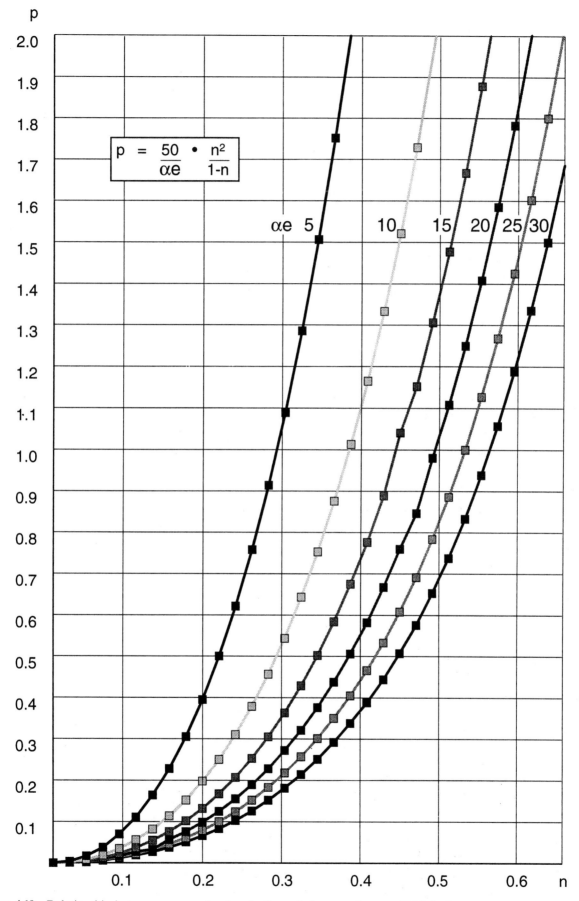

Figure 4.19 Relationship between p, α_e and n for elastic analysis; note that $p = 100A_s/b_w d$.

$$A_s = k_c k f_{ct.eff} A_{ct} / \sigma_s \qquad \text{(EC2 eqn. 4.78)}$$

This equation can be rewritten in the form

$$\sigma_s A_s = k_c k (f_{ct.eff} A_{ct}) \qquad (4.27)$$

Here A_s is the area of reinforcement within the tensile zone. σ_s is the maximum stress permitted in the reinforcement immediately after the formation of the crack. This value may be taken as $\boxed{100\%}$ of the yield strength of the reinforcement f_{yk}. It should be noted that a lower value may be needed to satisfy crack width limits; see section 4.8.3(c) below. $f_{ct.eff}$ is the tensile strength of the concrete effective at the time when the first cracks may be expected to occur. This is obviously difficult to quantify, and in EC2, it is suggested (cl. 4.4.2.2 P(3)) that when the time of cracking cannot be established with confidence as being less than 28 days, a minimum tensile strength of $\boxed{3}$ N/mm^2 may be adopted. k_c is a coefficient that takes account of the nature of the stress distribution immediately prior to cracking. k_c is taken as 1.0 for pure tension and 0.4 for bending. k is a coefficient that allows for the effect of non-uniform self-equilibrating stresses resulting from restraint to intrinsic and extrinsic deformations and varies between 0.5 and 1.0; see EC2 (cl. 4.4.2.2. P(3)).

In effect, equation (4.27) states that at the time of cracking the tensile capacity of the reinforcement in the tensile zone must be at least equal to the tensile capacity of the concrete. If it is assumed that the depth of the concrete tensile zone is approximately half the effective depth and A_s is replaced by pbd, then

$$p = (k_c k / 2) (f_{ct.eff} / \sigma_s)$$

With $\sigma_s = f_{yk} = 400$ N/mm^2, say $f_{ct.eff} = 3.0$ and $k_c k = 0.4$, then

$$p = 0.2 \times 3/400 = 0.0015$$

This agrees with the value given in section 4.6.3(a) above.

(c) Control of cracking without direct calculation (cl. 4.4.2.3)

A summary of the EC2 requirements for control of cracking is given below:

1. Cracking should be limited to a level that will not impair the proper functioning of the structure or cause its appearance to be unacceptable. Cracking is almost inevitable in reinforced concrete structures subject to bending, shear, torsion and tension resulting from either direct loading or restraint of imposed deformations. Cracks may also arise from other causes such as plastic shrinkage or expansive chemical reactions.

2. In the absence of specific requirements, it may be assumed that, for exposure classes 2–4, limitation of the maximum design crack width to about 0.3 mm under the quasi-permanent combination of loads will generally be satisfactory for reinforced concrete members in buildings with respect to appearance and durability.

3. For cracks caused dominantly by loading, their widths will not generally be excessive (that is, not greater than 0.3 mm) if either the provisions of Table 4.11 or Table 4.12 are complied with. Tables 4.10 and 4.11 have been modified from Tables 4.10 and 4.11 of EC2 to cover reinforced sections only. For spacing of stirrups in beams for crack control, see section 4.6.3(b) above. In Tables 4.11 and 4.12, the steel stresses should be evaluated for reinforced concrete on the basis of quasi-permanent loads.

(d) Calculation of crack widths (cl. 4.4.2.4)

P(1) The design crack width may be obtained from the relation

$$w_k = \beta \, s_{rm} \, \varepsilon_{sm} \qquad \text{(EC2 eqn. 4.80)}$$

where w_k is the design crack width, s_{rm} is the average final crack spacing, ε_{sm} is the mean strain allowing under the relevant combination of loads for the effects of tension stiffening, shrinkage, etc., and β is a coefficient relating the average crack width to the design value.

P(2) The values of β in equation (4.80) of EC2 may be taken as:

- $\beta = 1.7$ for load-induced cracking and for restraint cracking in sections with a minimum dimension in excess of 800 mm,
- $\beta = 1.3$ for restraint cracking in sections with a minimum dimension depth, breadth or thickness (whichever is the lesser) of 300 mm or less.

Values for intermediate section sizes may be interpolated. ε_{sm} may be calculated from the relation

$$\varepsilon_{sm} = \sigma_s / E_s [1 - \beta_1 \beta_2 (\sigma_{sr} / \sigma_s)^2]$$
$$\text{(EC2 eqn. 4.81)}$$

Table 4.10 Maximum bar diameters for high-bond bars (modified from table 4.10 of EC2)

Steel stress (MPa)	Maximum bar size (mm), reinforced sections
160	32
200	25
240	20
280	16
320	12
360	10
400	8
450	6

Table 4.11 Maximum bar spacings for high-bond bars (modified from table 4.11 of EC2)

Steel stress (MPa)	Maximum bar spacing (mm)	
	Pure flexure	Pure tension
160	300	200
200	250	150
240	200	125
280	150	75
320	100	–
360	50	–

where σ_s is the stress in the tension reinforcement calculated on the basis of a cracked section and σ_{sr} is the stress in the tension reinforcement calculated on the basis of a cracked section under the loading conditions causing first cracking. β_1 is a coefficient that takes account of the bond properties of the bars:

- $\beta_1 = 1.0$ for high-bond bars,
- $\beta_1 = 0.5$ for plain bars.

β_2 is a coefficient that takes account of the duration of the loading or of repeated loading:

- $\beta_2 = 1.0$ for a single, short-term loading,
- $\beta_2 = 0.5$ for a sustained load or for many cycles of repeated loading.

For members subjected only to intrinsic imposed deformations, σ_s may be taken as equal to σ_{sr}.

P(3) The average final crack spacing for members subjected dominantly to flexure or tension can be calculated from the equation

$$s_{rm} = 50 + 0.25\, k_1\, k_2\, \varphi/p_r \text{ (mm)}$$
(EC2 eqn. 4.82)

where φ is the bar size in mm. Where a mixture of bar sizes is used in a section, an average bar size may be used. k_1 is a coefficient that takes account of the bond properties of the bars:

- $k_1 = 0.8$ for high-bond bars,
- $k_1 = 1.6$ for plain bars.

k_2 is a coefficient that takes account of the form of the strain distribution:

- $k_2 = 0.5$ for bending,
- $k_2 = 1.0$ for pure tension.

For cases of eccentric tension or for local areas, intermediate values of k_2 should be used, which can be calculated from the relation

$$k_2 = (\varepsilon_1 + \varepsilon_2)/2\varepsilon_1$$

where ε_1 is the greater and ε_2 the lesser tensile strain at the boundaries of the section considered, assessed on the basis of a cracked section. p_r is the effective reinforcement ratio, $A_s/A_{c,eff}$, where A_s is the area of reinforcement contained within the effective tension area, $A_{c,eff}$. The effective tension area is generally the area of concrete surrounding the tension reinforcement of depth equal to 2.5 times the distance from the tension face of the section to the centroid of the reinforcement (see Figure 4.20). For slabs where the depth of the tension zone may be small, the height of the effective area should not be taken as greater than $(h-x)/3$.

(e) Control of deflection without direct calculation (cl. 4.4.3.1/2)

EC2 gives guidelines for limiting deflections in conventional buildings such as dwellings, offices, public buildings or factories. These are summarized as follows:

1. The appearance and general utility of a structure may be impaired when the calculated sag of a beam, slab or cantilever subject to quasi-permanent loads exceeds span/250. The sag is assessed relative to the supports. Precamber may be used to compensate for some or all of the deflection, but any upward deflection incorporated in the formwork should not generally exceed span/250.

2. Where partitions, etc., are in contact with or attached to members, it may be necessary to limit the deflection after construction to span/500.

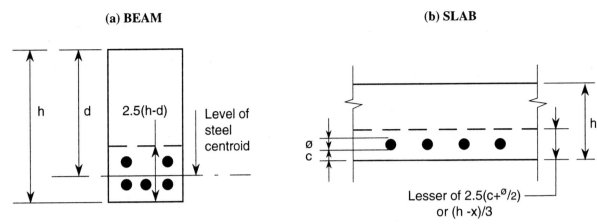

Figure 4.20 Effective tension areas for (a) beam and (b) slab.

Generally, adequate deflection control can be achieved by determining the limiting span/effective depth ratio from table 4.14 of EC2 and applying a number of correction factors. Table 4.14 of EC2 is reproduced in Table 4.12 in this chapter, and it should be noted that the span/effective depth values given are identical to those in the NAD, but the NAD introduces additional values for nominally reinforced members.

In particular, it must be emphasized that the values in Table 4.12 have been derived on the assumption that the steel stress, under the design service load at a cracked section at the midspan of a beam or slab or at the support of a cantilever, is 250 N/mm² (corresponding roughly to f_{yk} = 400 N/mm²). Where other stress levels are used, the values in Table 4.12 should be multiplied by $250/\sigma_s$ where σ_s is the stress at the section given above under the frequent combination of loads. It will normally be conservative to assume that

$$250/\sigma_s = 400/(f_{yk}A_{s,req}/A_{s,prov}) \qquad (4.28)$$

where $A_{s,prov}$ is the area of steel provided at the defined section and $A_{s,req}$ is the area of steel required at the section to give the required design ultimate moment of resistance. Further points to be noted are:

1. For flanged sections where the ratio of the flange breadth to the rib breadth exceeds 3, the values should be multiplied by 0.8.
2. For spans other than flat slabs exceeding 7 m, supporting partitions liable to be damaged by excessive deflections, the values should be multiplied by 7/span.
3. For flat slabs where the greater span exceeds 8.5 m, the values should be multiplied by 8.5/span.

4. Lightly stressed members are those where $p < 0.5\%$. This generally applies to slabs.
5. Highly stressed members correspond to $p = 1.5\%$, and for p values between 0.5 and 1.5, the span/effective depth ratio is obtained by interpolation; see Figure 4.21.

Consider an end span of a continuous beam of span 7.2 m with f_{yk} = 460 and $A_{s,req} = 0.9A_{s,prov}$ and $p = 1.25\%$. From Figure 4.21

basic L/d = 25.25

Modification for 7.2 m span is

$7/7.2 = 0.97$

Table 4.12 Basic ratios of span/effective depth for reinforced concrete members without axial compression (table 4.14 of EC2)

Structural system	Concrete highly stressed	Concrete lightly stressed
1. Simply supported beam, one- or two-way spanning simply supported slab	18	25
2. End span of continuous beam or one-way continuous slab or two-way spanning slab continuous over one long side	23	32
3. Interior span of beam or one-way or two-way spanning slab	25	35
4. Slab supported on columns without beams (flat slab) – based on longer span	21	30
5. Cantilever	7	10

(1) Simply supported beam, one-way or two-way spanning simply supported slab.

(2) End span of continuous beam or one-way continuous slab or two way spanning slab continuous over one long side.

(3) Interior span of beam or one-way or two-way spanning slab.

(4) Slab supported on columns without beams (flat slab) - based on longer span.

(5) Cantilever

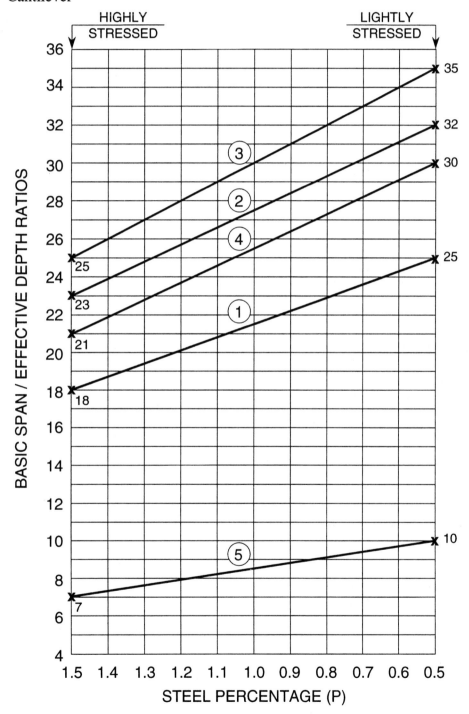

Figure 4.21 Relationship between steel percentage and basic span/effective depth ratio based on table 4.14 of EC2.

Modification for $A_{s,req} = 0.9A_{s,prov}$ and $f_{yk} = 460$ is

$$250/\sigma_s = 400/(460 \times 0.9) = 0.97$$

Thus

allowable $L/d = 25.25 \times 0.97 \times 0.97 = 23.76$

(f)   Estimation of deflection by calculation (EC2: Appendix 4)

Using elastic theory, the deflection δ of a beam can be expressed in the general form

$$\delta = kWL^3/EI$$

where W is the total load, L the span, EI the flexural rigidity and k a coefficient depending on the support conditions and type of loading. For a simply supported beam with a total uniformly distributed load W on a span L the bending moment is

$$M = WL/8$$

$$W = 8M/L$$

and

$$\delta = (5 \times 8/384)\ (M/EI)\ L^2 = 0.104\ (M/EI)\ L^2$$

The equation of bending gives

$$M/I = E/r = f/y$$

$$M/EI = 1/r$$

Thus

$$\delta = 0.104\ L^2\ (1/r)$$

In general,

$$\delta = kL^2\ (1/r) \tag{4.29}$$

For a continuous beam with uniformly distributed loading, it can be shown (Beckett, 1975) that for equal spans

$$k = 0.104\ (1 - \beta/10) \tag{4.30}$$

where

$$\beta = (M_A + M_B)/M_C$$

and M_A and M_B are the support moments and M_C is the midspan moment.

Thus calculation of deflections requires an estimation of curvature, and in EC2 A.4.3 a calculation method for curvature is presented as given below.

Two limiting conditions are assumed to exist for the deformation of concrete sections:

- The uncracked condition – the steel and concrete act together elastically in both tension and compression.
- The fully cracked condition – the influence of the concrete in tension is ignored.

If flexure dominates, an adequate prediction of behaviour is given by

$$\alpha = \xi\alpha_{II} + (1 - \xi)\alpha_I \qquad \text{(EC2 eqn. A.4.1)}$$

where α is the parameter considered, e.g. strain, curvature, rotation (and as a simplification, deflection). α_I and α_{II} are the values of the parameter calculated for the uncracked and fully cracked sections. ξ is a distribution coefficient given by

$$\xi = 1 - \beta_1\ \beta_2\ (\sigma_{sr}/\sigma_s)^2 \qquad \text{(EC2 eqn. A.4.2)}$$

where

$\beta_1 = 1$ for high-bond bars

$\beta_1 = 0.5$ for plain bars

β_2 is a coefficient that takes account of load duration,

$\beta_2 = 1$ for a single short-term loading

$\beta_2 = 0.5$ for sustained loading

σ_s is the stress in the tension steel calculated on the basis of a cracked section. σ_{sr} is the stress in the tension steel calculated on the basis of a cracked section under the loading which will just cause cracking at the section being considered (σ_s/σ_{sr} can be replaced by M/M_{cr} for flexure and N/N_{cr} for pure tension). ζ is zero for uncracked sections.

A best estimate of behaviour will be obtained if f_{ctm} is used for the tensile strength of concrete.

Creep may be allowed for using an effective modulus obtained from

$$E_{c,eff} = E_{cm}/(1 + \varphi) \qquad \text{(EC2 eqn. A.4.3)}$$

where φ is the creep coefficient (see table 3.3 of EC2).

Shrinkage curvature may be assessed from

$$1/r_{cs} = \varepsilon_{cs}\ \alpha_e\ S/I \qquad \text{(EC2 eqn. A.4.4)}$$

where $1/r_{cs}$ is the curvature due to shrinkage, ε_{cs} is the full shrinkage strain (see table 3.4 of EC2), S is the first moment of area of the reinforcement about the centroid of the section, I is the second moment of area of section and α_e is the effective modular ratio = $E_s/E_{c,eff}$. S and I should be calculated for the cracked and uncracked condition and the final curvature assessed from EC2 equation (A.4.1).

In buildings, it will normally be satisfactory to consider the deflections under the quasi-permanent combination of loading and assuming this load to be of long duration. EC2 requirements for controlling deflection and cracking with and without calculation are brought together in the following examples.

Example 4.6: Deflection check using table 4.14 of EC2

This example has been deliberately chosen to demonstrate the importance of a deflection check when the characteristic yield strength of the reinforcement (f_{yk}) exceeds 400 N/mm², the span effective depth ratio (L/d) is at the upper limit of table 4.14 of EC2 and a plastic analysis is adopted for the ultimate limit state. Consider a reinforced concrete slab continuous over two spans of 6.0 m (Figure 4.22) with design data as follows:

f_{ck} = 30 N/mm²

f_{yk} = 460 N/mm²

h = 220 mm

d = 190 mm $(L/d = 31.58)$

Q_k = 3.0 kN/m² (EC1 Draft, offices)

An allowance of 0.5 kN/m² is made for a suspended ceiling and services and the weight of the floor finishes is nominal (power floated surface and carpet). Light moveable partitions (say 0.5 kN/m²) will be treated as an imposed load with the partial safety factor appropriate to imposed loads $(\gamma_F = 1.5)$. Note that there are differences between BS 6399 and EC1 (Draft) with regard to partition loading and reference should be made to the forthcoming NAD for EC1.

Ultimate limit state

Using a plastic analysis for the ultimate limit state with equal support and span moments, the design moment is

$M_d = 0.086 \ (1.35 \ G_k + 1.5 \ Q_k) \ L^2$

$G_k = 0.22 \times 24 + 0.5 = 5.78 \text{ kN/m}^2$

$Q_k = 3.0 + 0.5 = 3.5 \text{ kN/m}^2$

For a 1.0 m width of slab

$M_d = 0.086 \ (1.35 \times 5.78 + 1.5 \times 3.5) \times 6^2$

$\quad = 0.086 \times 13.053 \times 6^2$

$\quad = 40.41 \text{ kN m}$

From Figure 4.3, with $f_{ck} = 30 \text{ N/mm}^2$,

$M/b_w d^2 = 40.41 \times 10^6/10^3 \times 190^2$

$\quad = 1.12$

$x/d \quad = 0.086 \quad (x/d < 0.25)$

$x \quad = 16.34 \text{ mm}$

Thus

$A_s = 40.41 \times 10^6/460/1.15 \ (190 - 0.4 \times 16.34)$

$\quad = 550.65 \text{ mm}^2/\text{m} \ (10 \ \varphi - 140 = 561 \text{ mm}^2/\text{m})$

Steel percentage

$p \quad = 100 \times 561/1000 \times 190 = 0.295$

Minimum steel area

$A_{s,min} = 0.6 \ b_t d/f_{yk}$

$\quad = 0.6 \times 10^3 \times 190/460$

$\quad = 247.8 \text{ mm}^2/\text{m}$

but not less than

$A_{s,min} = 0.0015 \ b_t d = 285 \text{ mm}^2/\text{m} \ (< 561)$

Check for shear at support B

$V = 13.053 \times 3.0 + 40.41/6 = 45.894 \text{ kN}$

Shear stress

$v = 45.894 \times 10^3/10^3 \times 190 = 0.242 \text{ N/mm}^2$

Basic shear stress for $f_{ck} = 30 \text{ N/mm}^2$ is $\tau_{Rd} = 0.34 \text{ N/mm}^2$; thus design adequate in shear.

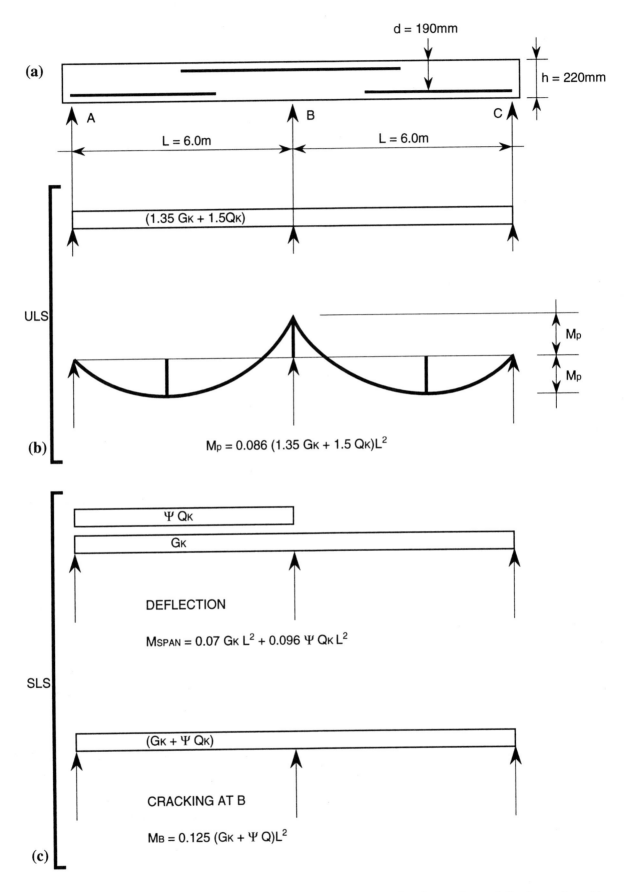

Figure 4.22 Examples 4.6 and 4.7: two-span reinforced concrete slab, loading for maximum deflection and cracking.

Deflection check

The slab can be classified as lightly stressed ($p <$ 0.5%) and thus from table 4.14 of EC2 the basic ratio of span/effective depth is 32. Note that the L/d values are related to a steel stress of 250 N/mm² corresponding roughly to f_{yk} = 400 N/mm². Where other stress levels are used, the values in table 4.14 should be multiplied by $250/\sigma_s$ where σ_s is the stress at the section under the frequent combination of loads. It will normally be conservative to assume that

$$250/\sigma_s = (400/f_{yk})(A_{s,req}/A_{s,prov})$$
$$= (400/460) \times (550.65/561)$$
$$= 0.884$$

Thus the L/d ratio should not exceed

$$(L/d)_{max} = 0.884 \times 32 = 28.3$$

The design appears to be unconservative in deflection. Noting that σ_s is the stress at the midspan of the slab under frequent combination of loads, its value can be estimated as follows.

The loading arrangement for maximum deflection is shown in Figure 4.22. The combination factor ψ_1 is taken as 0.5 (see Table 1.2) (0.6 in the NAD) under the frequent combinations of loads, and thus

$$G_k = 5.78 \text{ kN/m}$$

$$Q_k = 3.5 \times 0.5 = 1.75 \text{ kN/m}$$

From Figure 4.21, the midspan moment is

$$M_s = 0.07 \times 5.78 \times 6^2 + 0.096 \times 1.75 \times 6^2$$
$$= 14.56 + 6.05$$
$$= 20.61 \text{ kN m}$$

Under the rare combination of loads $\psi_0 = 0.7$ and thus

$$M_s = 14.56 + 6.05 \times 0.7/0.5$$
$$= 23.03 \text{ kN m}$$

From Table 1.6 the mean value of the tensile strength of the concrete is 2.9 N/mm² for f_{ck} = 30 N/mm². Thus the cracking moment is given by

$$M_{cr} = 2.9 \times 10^3 \times 220^2/6 \times 10^{-6}$$
$$= 23.39 \text{ kN m}$$

As this value approximates to M_s under the rare combination of loads, a cracked section will be assumed to estimate σ_s. As less than 50% of the

stress will arise from the quasi-permanent actions, long-term effects need not be taken into account and the modular ratio is given by

$$\alpha_e = E_s/E_{cm}$$

$$E_{cm} = 9.5 (f_{ck} + 8)^{1/3>} = 31.9 \text{ kN/mm}^2$$

Thus

$$\alpha_e = 200/31.9 = 6.27$$

For an elastic analysis the steel percentage is expressed (from equation (4.23)) as

$$p = (50/\alpha_e) \, n^2/(1 - n) \quad \text{(where } x = nd\text{)}$$

Alternatively, Figure 4.19 can be used. Thus

$$0.295 = (50/6.27) \, n^2/(1 - n)$$

and solving gives

$$n = 0.175$$

$$nd = 0.175 \times 190 = 33.25 \text{ mm}$$

and

$$\text{lever arm} = d - nd/3 = 178.9 \text{ mm}$$

Thus steel stress under the frequent combination of loads is

$$\sigma_s = 20.61 \times 10^6/561 \times 178.9$$
$$= 205.35 \text{ N/mm}^2$$

As σ_s is less than 250 N/mm², the L/d ratio of 32 can be adopted without modification and thus the design is adequate in deflection.

Example 4.7: cracking check using tables 4.11 and 4.12 of EC2

Using the design data for Example 4.6, a crack control check will be made for support B. The steel stresses used in tables 4.11 and 4.12 of EC2 should be evaluated for concrete on the basis of the quasi-permanent loads and thus $\psi_2 = 0.3$. The maximum bending moment at B (see Figure 4.22) is

$$M_B = 0.125 (G_k + \psi_2 \, Q_k) \, L^2$$
$$= 0.125 (5.78 + 0.3 \times 3.5)6^2$$
$$= 30.74 \text{ kN m}$$

Using the lever arm of 178.9 mm obtained in Example 4.6, the steel stress is

$$\sigma_s = 30.74 \times 10^6/561 \times 178.9$$

$$= 306.2 \text{ N/mm}^2$$

The reinforcement provided at support B is 10φ–140 c/c. Assuming the cracks are caused predominantly by loading, either the provisions of table 4.11 or the provisions of table 4.12 of EC2 should be complied with to avoid excessive crack widths. The provisions of table 4.11 are met and thus the steel provided for flexure at the ultimate limit state gives adequate crack control.

To comply with stress limitations (section 4.8.1 of this chapter) the rare combination loads will be used with $\psi_0 = 0.7$. This gives $M_B = 37.04$ kN m, $\sigma_s = 369$ N/mm^2 and $\sigma_c = 6.22$ N/mm^2. Thus the steel stress limitation of $\boxed{0.8}f_{yk}$ is just satisfied. The concrete stress is only $0.21f_{ck}$ and thus this does not present a serviceability problem.

Example 4.8: deflection check by calculation

The purpose of this example is to apply the procedure given in Appendix 4 of EC2 for estimating deflections by calculation. The Code states that, in buildings, it will normally be satisfactory to consider the deflections under quasi-permanent combination of loading, assuming this load to be of long duration. Consider a four-span continuous one-way slab (Figure 4.23) with the following design data:

- Four equal spans of 7.2 m
- $f_{ck} = 25$ N/mm^2
- $f_{yk} = 460$ N/mm^2
- Q_k (variable action) = 3.5 kN/m^2 (offices plus lightweight partitions)
- G_k (permanent action) based on unit weight of concrete plus 0.5 kN/m^2 for suspended ceiling and services
- $\psi_2 = 0.3$ quasi-permanent loading assumed long-term
- Cover to centroid of reinforcement, say 30 mm.

The following calculations relate to the end span.

Ultimate limit state

Loading
Assume a span/overall depth ratio of $L/h = 28$. Thus

$$h = 0.257 \text{ m, say, 260 mm}$$

$$G_k = 0.26 \times 24 + 0.5 = 6.74 \text{ kN/m}^2$$

$$Q_k = 3.5 \text{ kN/m}^2$$

Member analysis
Plastic analysis with equal support and span moments; thus the design moment and shear for the end span is

$$M_D = 0.086 \ (1.35 \times 6.74 + 1.5 \times 3.5) \times 7.2^2$$

$$= 63.97 \text{ kN m/m}$$

$$V_D = 14.35 \times 3.6 + 63.97/7.2$$

$$= 60.54 \text{ kN/m}$$

Section analysis
For *bending*

$$M/b_w d^2 = 63.97 \times 10^6/10^3 \times 230^2 = 1.21$$

From Figure 4.3

$$x/d = 0.11 \quad \text{for } f_{ck} = 25 \text{ N/mm}^2$$

$$x = 25.3 \text{ mm}$$

Thus

$$A_s = 63.97 \times 10^6/460/1.15 \times (230 - 0.4 \times 25.3)$$

$$= 727.3 \text{ mm}^2/\text{m}$$

Provide 12φ–150 (754 mm^2/m).

For *shear*, shear stress

$$v = V_D/b_w d$$

$$= 60.54 \times 10^3/10^3 \times 230$$

$$= 0.263 \text{ N/mm}^2$$

From Table 1.6 the basic shear strength τ_{Rd} is 0.3 N/mm^2 for $f_{ck} = 25$ N/mm^2; thus adequate in shear.

Serviceability limit state – deflection

For a continuous slab with uniformly distributed loading, the deflection in the end span can be expressed in the form

$$\delta = 0.104 \ (1 - \beta/10) \ L^2 \ (1/r) \qquad (4.30)$$

where

$$\beta = (M_A + M_B)/M_C$$

where M_A and M_B are the support moments and M_C is the midspan moment. The loading arrangement is shown in Figure 4.23:

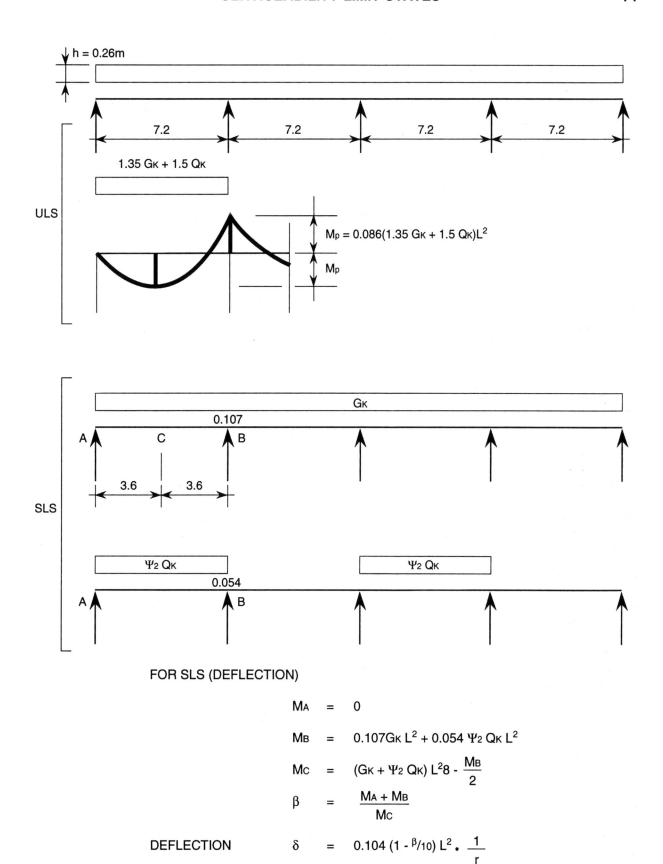

$$M_A = 0$$

$$M_B = 0.107 G_K L^2 + 0.054 \, \Psi_2 \, Q_K \, L^2$$

$$M_C = (G_K + \Psi_2 \, Q_K) \, L^2 8 - \frac{M_B}{2}$$

$$\beta = \frac{M_A + M_B}{M_C}$$

DEFLECTION $\quad \delta = 0.104 \, (1 - {}^\beta/10) \, L^2 \cdot \dfrac{1}{r}$

Figure 4.23 Examples 4.8 and 4.9: four-span slab, loading arrangements.

$M_A = 0$ (simple end support assumed)

$$M_B = 0.107\, G_k L^2 + 0.054\, \psi_2\, Q_k\, L^2$$
$$= 0.107 \times 6.74 \times 7.2^2 + 0.054 \times 0.3 \times 3.5$$
$$\times 7.2^2$$
$$= 37.39 + 2.94$$
$$= 40.33 \text{ kN m/m}$$

$$MC = (G_k + \psi_2\, Q_k)\, L^2/8 - M_B/2$$
$$= (6.74 + 0.3 \times 3.5)\, 7.2^2/8 - 40.39/2$$
$$= 50.48 - 20.20$$
$$= 30.28 \text{ kN m}$$

$$\beta = (M_A + M_B)/M_c = 40.33/30.28 = 1.332$$

$$1 - \beta/10 = 0.867$$

$$k = 0.104 \times 0.867 = 0.090$$

Thus deflection

$$\delta = 0.090\, L^2\, (1/r)$$

The final stage of the calculation is to estimate the curvature from the deformation parameter (curvature) α

$$\alpha = \zeta \alpha_{II} + (1 - \zeta)\alpha_I$$

noting that α_I and α_{II} are the parameters for the uncracked and fully cracked sections respectively.

The distribution factor ζ is obtained from

$$\zeta = 1 - \beta_1\, \beta_2\, (M/M_{cr})^2$$

The coefficient for bond β_1 will be taken as 1.0 (high-bond bars) and the duration of loading coefficient β_2 will be taken as 0.5 (sustained loading). The cracking moment

$$M_{cr} = f_{ctm} b_w h^2/6$$

For $f_{ck} = 25$ N/mm^2, $f_{ctm} = 2.6$ N/mm^2. Thus

$$M_{cr} = 2.6 \times 10^3 \times 260^2/6 \times 10^{-6} = 29.29 \text{ kN m/m}$$

The midspan moment was estimated to be 30.28 kN m/m. Thus

$$M/M_C = 30.28/29.29 = 1.033$$

Thus

$$\zeta = 1 - 1.0 \times 0.5 \times 1.033^2 = 0.466$$

To determine the value of α_I, the equation of bending gives

$$\alpha_I = 1/r = M/EI$$

For $f_{ck} = 25$ N/mm^2, E_{cm} is obtained from

$$E_{cm} = 9.5\, (f_{ck} + 8)^{1/3} = 30.4 \text{ kN/mm}^2$$

Creep is allowed for using an effective modulus of

$$E_{c,eff} = E_{cm}/(1 + \varphi)$$

Taking φ as 2.0, say (see table 3.3 of EC2),

$$E_{c,eff} = 30.4 \times 0.33 = 10.14 \text{ kN/mm}^2$$

Thus

$$\alpha_I = 1/r = 29.29 \times 10^6 \times 12/10.14 \times 10^3$$
$$\times 10^3 \times 260^3$$
$$= 1.972 \times 10^{-6}$$

The value of α_{II} is obtained (see Figure 4.24) from

$$1/r = \sigma_s/E_s y$$

The steel percentage

$$p = 754/10^3 \times 230 \times 10^3 = 0.33$$

The neutral axis factor n is obtained by solving the quadratic $p = (50/\alpha_e)n^2/(1 - n)$ or from Figure 4.19. Using the effective modulus $E_{c,eff} = 10.14$ kN/mm^2

$$\alpha_e = E_s/E_{c,eff} = 200/10.14 = 19.72 \text{ (say, 20)}$$

From Figure 4.19 with $p = 0.33$ and $\alpha_e = 20$

$$n = 0.36 \text{ (approx.)}$$

$$d(1 - n/3) = 202.4$$

Thus steel stress

$$\sigma_s = 30.28 \times 10^6/754 \times 202.4 = 198 \text{ N/mm}^2$$

$$d(1 - n) = y = 147.2 \text{ mm}$$

and

$$\alpha_{II} = 1/r = \sigma_s/E_s y$$
$$= 198/200 \times 10^3 \times 147.2$$
$$= 6.725 \times 10^{-6}$$

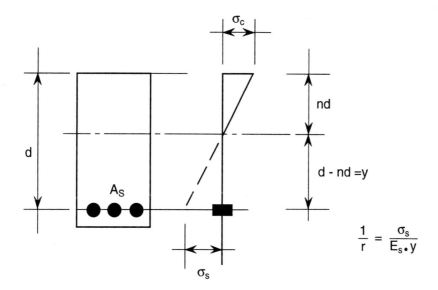

Figure 4.24 Example 4.8: diagram to obtain $1/r$.

Thus

$$\alpha = [0.466 \times 6.725 + (1 - 0.466) \times 1.972]10^{-6}$$

$$= 4.187 \times 10^{-6}$$

Thus deflection

$$\delta = 0.09 \, L_2 \, (1/r)$$

$$= 0.09 \times 7.2^2 \times 10^6 \times 4.187 \times 10^{-6}$$

$$= 19.53 \text{ mm}$$

$$L/\delta = 7200/19.53 = 369$$

This value is well within the guideline of span/250 relevant to the appearance and general utility of the structure.

Example 4.9: crack width by calculation

Using the design data for Example 4.8, the crack width will be estimated at support B. The loading conditions are shown in Figure 4.25. The first stage is to calculate the stress σ_s in the tension reinforcement on the basis of a cracked section and the stress σ_{sr} in the tension reinforcement calculated on the basis of a cracked section under the loading conditions causing first cracking. The moment at B for $G_k + \psi_2 \, Q_k$ is given by (see Figure 4.25):

$$M_B = 0.107 \, G_k \, L^2 + 0.121 \, \psi_2 \, Q_k \, L^2$$

$$= 0.107 \times 6.74 \times 7.2^2 + 0.121 \times 0.3$$

$$\times 3.5 \times 7.2^2$$

$$= 37.39 + 6.59$$

$$= 43.98 \text{ kN m/m}$$

From Example 4.8, $p = 0.33$ and using the long-term value of $\alpha_e = 20$, the neutral axis factor from Figure 4.19 is $n = 0.36$. Thus

$$\sigma_s = M_B/A_s \, d \, (1 - n/3)$$

$$= 43.98 \times 10^6/754 \times 202.4$$

$$= 288.2 \text{ N/mm}^2 < 0.8 \, f_{yk} = 368$$

The cracking moment is 29.29 kN m (see Example 4.8). Thus

$$\sigma_{sr} = 29.29 \times 10^6/754 \times 202.4$$

$$= 191.9 \text{ N/mm}^2$$

$$(\sigma_{sr}/\sigma_s)^2 = 0.443$$

The mean strain

$$\varepsilon_{sm} = \sigma_s/E_s[1 - \beta_1 \, \beta_2 \, (\sigma_{sr}/\sigma_s)^2]$$

Taking $\beta_1 = 1.0$ for high-bond bars and $\beta_2 = 0.5$ for sustained loading,

$$\varepsilon_{sm} = 288.2/200 \times 10^3 \, (1 - 0.5 \times 0.443)$$

$$= 1.12 \times 10^{-3}$$

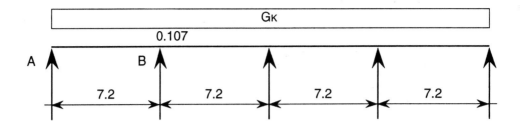

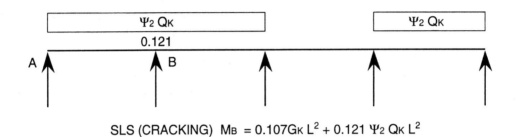

SLS (CRACKING) $M_B = 0.107G_K L^2 + 0.121 \Psi_2 Q_K L^2$

Figure 4.25   Example 4.9: loading to estimate crack width.

The average final crack spacing is calculated from

$$s_{rm} = 50 + 0.25 \, k_1 \, k_2/p_r$$

$$\varphi = \text{bar diameter} = 12 \text{ mm}$$

$$k_1 = 0.8 \text{ (high-bond bars)}$$

$$k_2 = 0.5 \text{ (bending)}$$

For slabs, the height of the effective tension area should not exceed $(h - x)/3$. From

$$x = n \, d = 0.36 \times 230 = 82.8$$

we get

$$(h - x)/3 = 59.07 \text{ mm}$$

$$2.5 \times 30 = 75$$

Thus

$$A_{c,eff} = 49.07 \times 10^3$$

$$A_s = 754$$

Thus

$$p_r = 754/59.07 \times 10^3 = 12.76 \times 10^{-3}$$

The design crack width is given by

$$W_k = \beta \, s_{rm} \, \varepsilon_{sm}$$

$$\beta = 1.7 \quad \text{(load-induced cracking)}$$

$$s_{rm} = 50 + 0.25 \times 0.8 \times 0.5 \times 12/12.76 \times 10^{-3}$$
$$= 50 + 94.04$$
$$= 144.04 \text{ mm}$$

$$\varepsilon_{sm} = 1.12 \times 10^{-3}$$

Thus

$$W_k = 1.7 \times 144.04 \times 1.12 \times 10^{-3} = 0.274 \text{ mm}$$

This is within the limitation of a maximum design crack width of about 0.3 mm under the quasi-permanent combination of loads.

4.9 SUMMARY

4.9.1 Flexure and shear at ultimate limit state

Section analysis for flexure and shear at the ultimate limit state in accordance with EC2 is straightforward. The procedure is summarized below for a given design bending moment M_{sd} and shear force V_d:

1. Use the design chart, Figure 4.3, to determine the value of x, noting the upper limits of x/d.
2. The steel area A_s for flexure is obtained from the equation

$$M_{sd} = A_s f_{yk}/\gamma_s \, (d - 0.4\, x)$$

3. For a known reinforcement ratio $p_1 = A_{s1}/b_w d$, determine V_{Rd1} from equation (4.2), that is,

$$V_{Rd1} = \tau_{Rd}\, k\, (1.2 + 40\, p_1)\, b_w d$$

4. If $V_{Rd1} < V_{sd}$, check that V_{Rd2} (from Table 4.2) $\geqslant V_{sd}$ and then determine $V_{wd} = V_{sd} - V_{Rd1}$.
5. Use equation (4.5) to determine the spacing and cross-sectional area of vertical links, that is,

$$V_{wd} = (A_{sw}/s) \times 0.9\, d f_{ywd}$$

6. Check that the design complies with EC2 detailing provisions, which are summarized in section 4.6 of this chapter.

4.9.2 Torsion at ultimate limit state

In conventional reinforced concrete frameworks it is generally possible to arrange the structural elements such that it will not be necessary to consider torsion at the ultimate limit state. However, the possibility of torsions arising from considerations of compatibility should not be ignored; see section 4.5. In the event of the need to quantify torsion reinforcement by calculation, the procedure given in Example 4.3 should be followed.

4.9.3 Design of beams for seismic actions

It should be noted that EC8 considers three levels of ductility – low, medium and high. Example 4.4 demonstrates how EC2 is used in conjunction with EC8 and deals with the influence of ductility levels on reinforcement requirements.

4.9.4 Serviceability limit states

Considerable attention has been paid in this chapter to serviceability limit states. It can be seen from Examples 4.6 to 4.9 that estimation of crack width and deflection by direct hand calculation is, to say the least, tedious and in general can be avoided. Use should be made of tables 4.11 to 4.14 of EC2 to check crack control and deflection without direct calculation, noting the use of different load combinations – rare (ψ_0), frequent (ψ_1) and quasi-permanent (ψ_2).

REFERENCES

BCA (1990) *British Cement Association, Ove Arup, Square Grip, Flat Slab Shear Reinforcement Manual (the development of a range of prefabricated shear reinforcement systems for flat slab construction)*, British Cement Association.

Beckett D. (1975) Plastic analysis of continuous reinforced concrete beams, MPhil thesis, University of Surrey.

Kong F.K. and Evans R.H. (1987) *Reinforced and Prestressed Concrete*, 3rd edn, Van Nostrand Reinhold, New York.

Narayanan, R. (1986) Design of concrete structures for torsion (chapter 5), *Concrete Framed Structures, Stability and Strength* (ed. R. Narayanan), Elsevier Applied Science, London

Regan P.E. (1986) Design of reinforced concrete flat slabs (chapter 8), *Concrete Framed Structures, Stability and Strength* (ed. R. Narayanan), Elsevier Applied Science, London.

5

SECTION ANALYSIS (2): BEAM–COLUMN JOINTS

5.1 INTRODUCTION

In the design of low/medium-rise reinforced concrete frameworks for vertical and wind loading, the magnitude of the shear forces induced in the joints will not, in general, be at a level at which detailed design provisions are necessary. However, if the framework is subjected to seismic actions, including cyclic reversals of actions, then the shear forces induced can be an order of magnitude greater than those for wind loading. EC8 gives detailed design provisions for beam–column joints at the three ductility levels, DC 'H', DC 'M' and DC 'L'. Since the publication of the 1988 draft of EC8, the results of a comprehensive research programme (undertaken in New Zealand, Japan, China and the USA), involving testing of full-scale beam–column–slab joint assemblies under quasi-static cyclic loading, have been published (ACI, 1991). Reference is made to this research programme in a recent paper by Cheung *et al.* (1993), and design criteria, behavioural models, joint shear strength and anchorage of beam bars within joint cores are also discussed. The EC8 and New Zealand Code design provisions for beam–column joints are broadly similar, that is: 'the design criteria are intended to ensure that the strength of a beam–column joint core should not be less than that corresponding with the development of the selected plastic hinge mechanism in the frame and that the capacity of the column should not be jeopardised by possible strength degradation of the joint' (Cheung *et al.*, 1993). The results of the international collaborative research report (ACI, 1991) are summarized below:

1. Strong column–weak beam behaviour (plastic hinges are designed to form in the beams rather than the columns, thus the columns above and below a beam–column joint remain elastic) is desirable and can be achieved.
2. The influence of column axial load has little influence on ultimate shear strength of joints.
3. Limits are proposed for joint shear strain.
4. There is advantage in relocating beam plastic hinge rotation away from the column force, but sliding shear deformation may occur at the plastic hinge.
5. Design recommendations are given for the effective width of tension flanges.
6. The maximum effective beam width should be twice the column width.
7. Eccentric beam–column joints cause large torsional moments in columns.
8. Recommendations are given for increased anchorage lengths in beam–column joints.

Reports on the Erzincan earthquake in Turkey on 13 March 1993 (NCE, 1993) have confirmed the importance of beam–column joint design, as it was a primary cause of failures, in particular, lack of confinement steel (no horizontal links in the joint core) and poor detailing in general. The three references cited in this chapter are essential reading where building design involves seismic actions.

5.2 JOINT BEHAVIOUR

The deflected form and associated bending moment diagram for an internal joint of a framework are

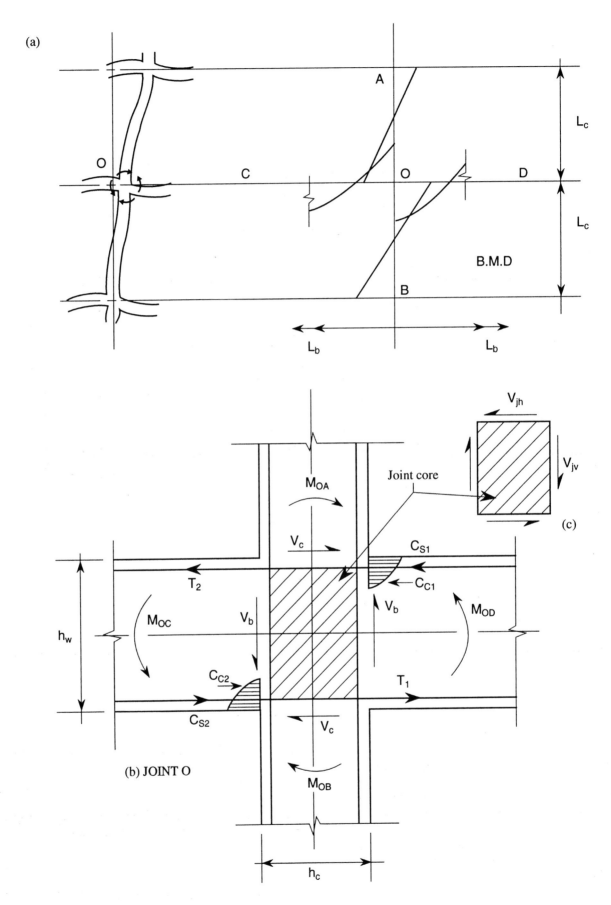

Figure 5.1 Actions on a beam–column joint and joint core.

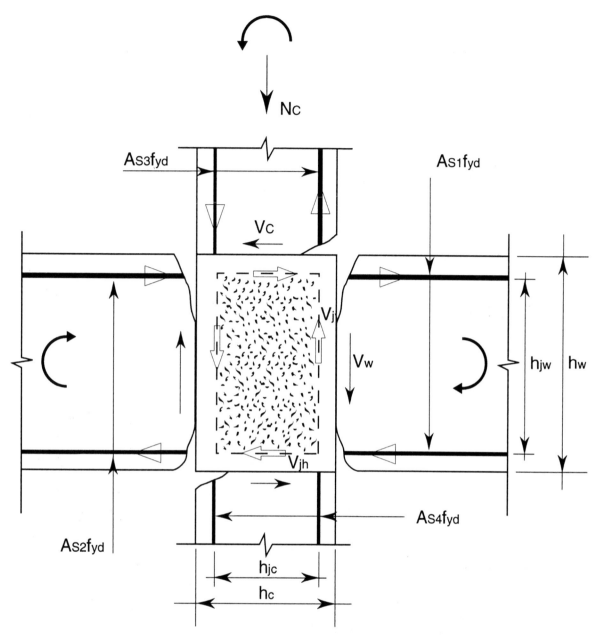

Figure 5.2 Adaptation of EC8 example of simplified determination of shear forces acting on the concrete core of a joint; equations for V_{jh} and V_{jv} are given in the text (equations (5.3)–(5.5)).

shown in Figure 5.1 (a). The forces acting on the joint are shown in Figure 5.1(b) and the joint core shears in Figure 5.1(c). The following simplified approach after Cheung *et al.* (1993) assumes that for building frameworks of regular configuration the seismic shear forces from the beams at the opposite sides of the joint core are similar and equal to V_b and those for the columns equal to V_c. The beam and column shears arise from the change in bending moment over their lengths and heights respectively. Thus:

$$V_c \sim (M_{OA} + M_{AO})/L_c$$

$$V_B \sim (M_{OC} + M_{CO})/L_b$$

Noting that $T_2 = C_{C2} + C_{S2}$ and $T_1 = C_{C1} + C_{S1}$, then

$$V_{jh} = T_1 + T_2 - V_c \qquad (5.1)$$

Ignoring the influence of axial load, the vertical shear on the joint core may be approximated as

$$V_{jv} = V_{jh}\, h_w/h_c \qquad (5.2)$$

EC8 adopts a similar approach to the above, and Figure 5.2 is an adaptation of an example given in the Code of a simplified determination of the shear forces acting on the core of a joint. The difference in notation and sense of the applied moments should be noted. The values of V_{jh} and V_{jv} are

$$V_{jh} = \gamma_{Rd} \left[2/3 \ (A_{s1} + A_{s2} \ q/5) \ f_{yd} \right] - V_c \qquad (5.3)$$

$$V_{jv} = \gamma_{Rd} \left[2/3 \ (A_{s3} + A_{s4}) \ f_{yd} \right] - V_w + N_c/2 \qquad (5.4)$$

(the factor $q/5$ was introduced in the 1993 draft of EC8), or, as a simplification

$$V_{jv} = V_{jh} \ h_w/h_c$$

It can be seen that equation (5.1) is similar to equation (5.3) and equation (5.2) is identical to equation (5.4), simplified version. The factor γ_{Rd} is introduced to balance the γ_s value ($f_{yd} = f_{yk}/\gamma_s$) and to compensate for strain hardening of the reinforcement. The reduction factor of 2/3 is to allow for part of the inclined bond forces flowing out of the core of the joint. Conventionally, the shear force transfer across a joint core can be effected by two mechanisms. These are shown in Figure 5.3. In EC8, these are referred to as the diagonal strut (a) and trusses and struts (b).

5.2.1 Diagonal struts

In the mechanism of Figure 5.3(a) it is assumed that narrow flexural cracks at the beam ends, caused by previous reversal of moderate seismic actions, are subsequently *closed*. Horizontal compressive forces

are transferred through the concrete compression zone and are combined with the vertical forces of the compressed zone of the column. Thus a diagonal compressive strut is formed, self-equilibrated within the joint. In this case, it is assumed that the compressive strength of the concrete, under simultaneous transverse tension, is governed by the bearing capacity of the joint. In EC8, the integrity of the diagonal strut is assumed to be maintained if

$$V_{jh} \leq 20\tau_{Rd}b_jh_c \qquad \text{for interior joints} \qquad (5.5)$$

$$V_{jh} \leq 15\tau_{Rd}b_jh_c \qquad \text{for exterior joints}$$

where b_j is the effective joint width as defined in Figure 5.4.

5.2.2 Trusses and struts

In this mechanism (Figure 5.3 (b)) it is assumed that if wide flexural cracks at beam ends, caused by previous reversal of major seismic actions, are *not closed*, the horizontal compressive forces may be transferred only through reinforcement of the beam. It is assumed that a complete diagonal strut cannot develop and that there is penetration of yield of the reinforcement into the joint, resulting in high bond stresses. Thus diagonal cracks within the core of the joint cannot be avoided. Thus an additional mechanism is necessary for shear transfer requiring vertical and horizontal reinforcement. With the provision of this reinforcement, EC8 allows the maximum tensile stress in the concrete to be limited to

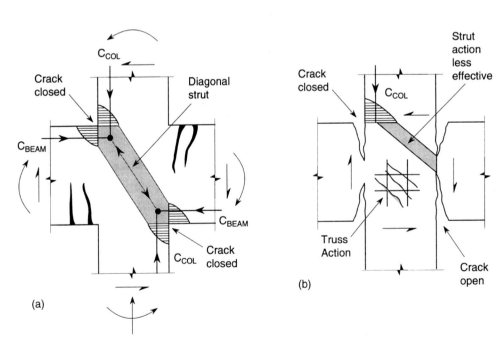

Figure 5.3 Mechanisms for effecting shear force transfer across a joint core: (a) diagonal strut; (b) trusses and struts.

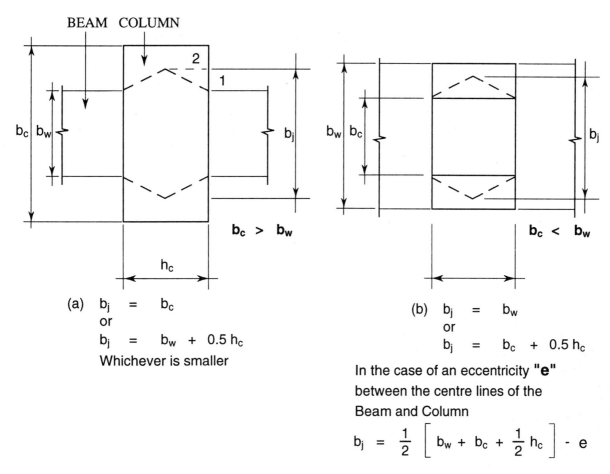

Figure 5.4 Effective joint width, b_j, in two different cases of beam and column widths: (a) $b_c > b_w$; (b) $b_c < b_w$.

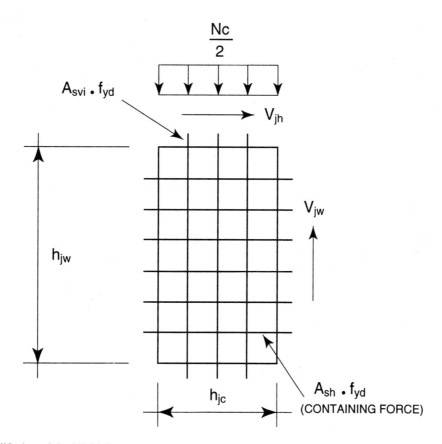

Figure 5.5 Simplified model of EC8 for mean values of shear and normal forces acting on a concrete joint; these values are given in equations (5.7)–(5.9) in the text.

$$\sigma_{ct} \leqslant f_{ctm}/\gamma_c \qquad (5.6)$$

where f_{ctm} is the mean concrete tensile strength; see Table 1.6.

To determine the value of σ_{ct}, EC8 (1988) gives a simplified model, with notation as in Figure 5.5. Mean values of the shear and normal stresses acting on the concrete joint may be obtained from the following equations:

$$\tau_h = [\gamma_{Rd} \times 2/3 \ (A_{s1} + A_{s2}) \ f_{yd} - V_c]/b_j h_{jw} \qquad (5.7)$$

$$\sigma_v = (N_c/2 + A_{svi}\lambda f_{yd})/b_j h_{jc} \qquad (5.8)$$

$$\sigma_h = A_{sh} \ f_{yd}/b_j h_{jw} \qquad (5.9)$$

In equation (5.8), λ allows for the precompression of the longitudinal bars in the column. The principal tensile stress σ_{ct} (see Figure 5.6) is obtained from

$$\sigma_{ct} = (\sigma_h + \sigma_v)/2 \pm \{[(\sigma_h - \sigma_v)/2]^2 + \tau_h^2\}^{1/2} \quad (5.10)$$

The application of the above equations to joint design is demonstrated in Example 5.1; see later. As with beam design for seismic actions, the complexity of joint design increases with increase in ductility requirements.

In the 1993 draft of EC8, it is stated that the diagonal strut method (mechanism (a)) shall be favoured and the trusses and struts method (mechanism (b)) shall be avoided. In the absence of more precise data, this may be satisfied by adopting the procedures given in sections 5.4 and 5.5.

5.3 EC8 BEAM–COLUMN JOINTS, DUCTILITY CLASS 'L'

For DC 'L' beam–column joints, calculation of horizontal confinement reinforcement is not required and the following rules apply:

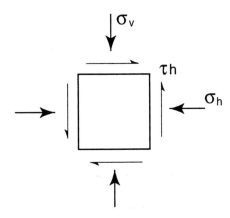

Figure 5.6 Principal tensile stress; the formula for σ_{ct} is given in the text by equation (5.10).

1. The horizontal confinement reinforcement in beam–column joints shall be equal to that provided at critical regions of the column (see Chapter 6).
2. At least one intermediate vertical bar is provided between column corners at each side of the joint.

5.4 EC8 BEAM–COLUMN JOINTS, DUCTILITY CLASS 'M'

5.4.1 Joint shear forces

The joint core shear forces are determined under the most adverse conditions of seismic loading, that is, capacity design conditions for the concurring beam ends ($\gamma_{Rd} = 1.15$) and the lowest compatible values of axial and shear forces from the framing columns.

5.4.2 Diagonal strut

The integrity of the diagonal strut is obtained from equation (5.5), that is

$$V_{jh} \leqslant 20 \ \tau_{Rd}b_j h_c \quad \text{interior joints}$$

$$V_{jh} \leqslant 15 \ \tau_{Rd}b_j h_c \quad \text{exterior joints}$$

Values of τ_{Rd} are obtained from Table 1.6 and b_j and h_c are defined in Figure 5.4.

5.4.3 Joint confinement

Adequate confinement (both vertical and horizontal) of the joint shall be provided to reduce the maximum diagonal tensile stress of the concrete so that

$$\sigma_{ct(max)} \leqslant f_{ctm}/\gamma_c$$

In the absence of a more precise model, the above may be satisfied if the following hold:

1. Adequate horizontal links (hoops) shall be provided within the joint so that

$$A_{sh} \ f_{yd}/b_j \ h_{jw} \geqslant V_{jh}/b_j \ h_{jc}$$
$$- \lambda \ [\tau_{Rd} \ (12\tau_{Rd} + v_d f_{cd})]^{0.5} \qquad (5.11)$$

where τ_{Rd} is obtained from Table 1.6, λ is a factor accounting for the available shear resistance of plain concrete after cyclic degradation

(λ = 1.2 for DC 'M'), v_d (= $N_c/A_c f_{cd}$) is the normalized design axial force with N_c under the combination considered, and V_{jh} is obtained from equation (5.3) with γ_{Rd} = 1.15.

2. Adequate vertical reinforcement of the column passing through the joint shall be provided so that

$$A_{sv,i} \geq 2/3 \, A_{sh} h_{jc}/h_{jw} \qquad (5.12)$$

Note that equation (5.11) is written in terms of equality of stresses. Equation (5.11) can be rewritten in a simplified form as follows (in terms of equality of forces, not stresses):

$$V_{sh} \geq V_{jh} - V_{ch} \qquad (5.13)$$

In equation (5.13), V_{sh} represents the contribution of the horizontal confinement reinforcement and V_{ch} the contribution of the diagonal compression struts. A similar expression is also given by Cheung *et al.* (1993).

5.4.4 Reinforcement

The minimum horizontal confinement reinforcement to be provided is 6 mm φ with a spacing the lesser of $h_c/2$ or 150 mm. Additionally, at least one intermediate vertical bar is provided between column corners on each side of the joint.

5.5 EC8 BEAM–COLUMN JOINTS, DUCTILITY CLASS 'H'

5.5.1 Joint shear forces

The joint shear forces are determined as in section 5.4.1 but with γ_{Rd} = 1.25.

5.5.2 Diagonal strut

Procedure as in section 5.4.2.

5.5.3 Joint confinement

Procedure as in section 5.4.3, but in equation (5.11), γ_{Rd} = 1.25 and the factor λ = 1.0.

5.5.4 Reinforcement

The minimum horizontal confinement reinforcement to be provided is 6 mm φ with a spacing the lesser

of $h_c/4$ or 100 mm. If framing beams are on all four faces of the column, the spacing of the horizontal confinement hoops may be reduced to $h_c/2$, but not greater than 150 mm. At least one intermediate vertical bar is provided between column corners, but the maximum distance between consecutive bars is limited to 150 mm. The requirements for beam–column joint design are brought together in the following example.

Example 5.1: beam–column joint, DC 'H'

As the design requirements for DC 'L' are nominal and those for DC 'M' and DC 'H' are similar, this example will consider DC 'H' only. The joint geometry is shown in Figure 5.7 with 550×250 mm^2 beams framing into the four sides of the 400×400 mm column. The design data are as follows:

$$f_{ck} = 30 \text{ N/mm}^2 \qquad \tau_{Rd} = 0.34 \text{ N/mm}^2$$

$$f_{yk} = 400 \text{ N/mm}^2 \qquad q = 5 \qquad \gamma_{Rd} = 1.25$$

For actions in the X direction:

$$A_{s1} + A_{s2} = 804 \text{ mm}^2 \qquad (4\text{--}16\varphi)$$

$$A_{s3} + A_{s4} = 1610 \text{ mm}^2 \qquad (8\text{--}16\varphi)$$

with V_c = 28 kN, V_w = 34 kN and N_c = 280 kN. Thus from equation (5.3)

$$V_{jh} = \gamma_{Rd} \left[2/3 \, (A_{s1} + A_{s2}) f_{yd} \right] - V_c$$

$$\text{(note } q/5 = 1.0)$$

$$= 1.25 \times (2/3) \times 804 \times (400/1.15) \times 10^{-3} - 28$$

$$= 233 - 28$$

$$= 205 \text{ kN}$$

From equation (5.4)

$$V_{jv} = \gamma_{Rd} \left[2/3 \, (A_{s3} + A_{s4}) f_{yd} \right] + N_c/2 - V_w$$

$$= 1.25 \times (2/3) \times 1610 \times (400/1.15)$$

$$\times 10^{-3} + 280/2 - 34$$

$$= 467 + 140 - 34$$

$$= 573 \text{ kN}$$

For actions in the Y direction:

$$A_{s1} + A_{s2} = 1610 \text{ mm}^2 \qquad (8\text{--}16\varphi)$$

$$A_{s3} + A_{s4} = 2510 \text{ mm}^2 \qquad (8\text{--}20\varphi)$$

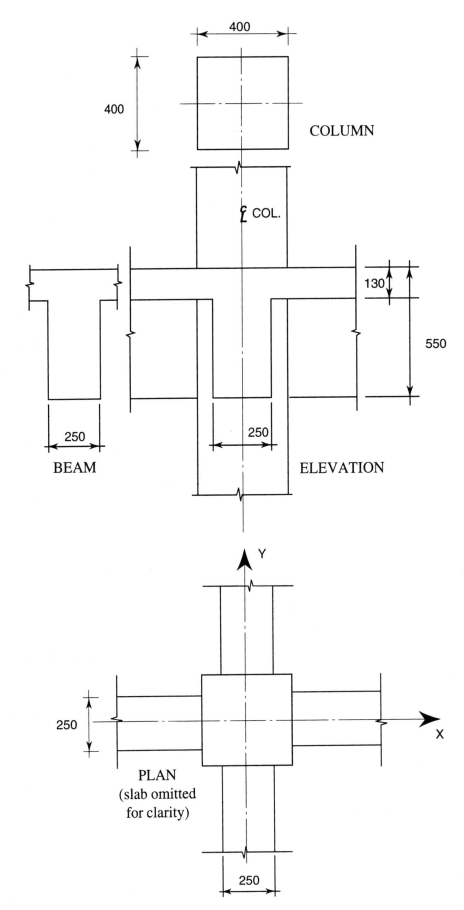

Figure 5.7 Example 5.1: details of the sizes of the members for a beam–column joint designed to DC 'H'.

with $V_c = 18$ kN, $V_w = 16$ kN and $N_c = 480$ kN. Thus, as actions in the X direction

$$V_{jh} = 1.25 \times (2/3) \times 1610 \times (400/1.15)$$
$$\times 10^{-3} - 18$$
$$= 467 - 18$$
$$= 449 \text{ kN}$$

$$V_{jv} = 1.25 \times (2/3) \times 2510 \times (400/1.15)$$
$$\times 10^{-3} + 480/2 - 16$$
$$= 728 + 240 - 16$$
$$= 952 \text{ kN}$$

In EC8 there is an approximation to V_{jv} as $V_{jv} = V_{jh}(h_w/h_c)$.

The values of V_{jh} are checked against the integrity of the diagonal strut (equation (5.5)), that is

$$V_{jh} \leqslant 20\tau_{Rd}b_jh_c \quad \text{(interior joint)}$$

Thus $b_j = b_c = 400$, $\tau_{Rd} = 0.34$ N/mm^2 ($f_{ck} = 30$ N/mm^2), $h_c = 400$ and b_j is the lesser of

$$b_c = 400 \text{ mm}$$

or

$$b_w + h_c/2 = 250 + 400/2 = 450 \text{ mm}$$

Thus $b_j = b_c = 400$ mm and

$$V_{jh} = 20 \times 0.34 \times 400 \times 400 \times 10^{-3}$$
$$= 1088 > 852 \text{ kN}$$

Thus the diagonal strut integrity requirement is met.

The required horizontal reinforcement is obtained from equation (5.12) with the factor 1.2 omitted. The components of equation (5.12) are evaluated below for direction Y with $(A_{s1} + A_{s2}) = 1610$ mm^2, $V_c = 18$ kN, $b_j = 400$ mm and $h_{jc} = 300$ (say) and $h_{jw} = 450$ (say):

$$[\tfrac{2}{3} \gamma_{Rd} (A_{s1} + A_{s2}) f_{yd} - V_c]/b_jh_{jc}$$
$$= [\tfrac{2}{3} \times 1.25 \times 1610 \times (400/1.15)$$
$$- 18 \times 10^3]/400 \times 300$$
$$= (466\,666 - 18\,000)/400 \times 300$$
$$= 3.74 \text{ N/mm}^2$$

$$V_d = N_c/A_cf_{cd}$$
$$= 480 \times 10^3/[400^2 \times (30/1.5)]$$
$$= 0.15 \text{ N/mm}^2$$

Thus

$$[\tau_{Rd} (12\ \tau_{Rd} + v_d\ f_{cd})]^{1/2}$$
$$= [0.34 (12 \times 0.34 + 0.15 \times 20)]^{1/2}$$
$$= 1.55 \text{ N/mm}^2$$

Hence

$$A_{sh}f_{yd}/b_j\ h_{jw} = 3.74 - 1.55 = 2.15 \text{ N/mm}^2$$

Thus

$$A_{sh} = 2.19 \times 400 \times 450/(400/1.15) = 1133 \text{ mm}^2$$

and

$$A_{svi} = 1133 \times 300/450 \times 2/3 = 503 \text{ mm}^2$$

The values of A_{sh} and A_{svi} above must be at least equal to the minimum requirements for DC 'H'; see section 5.5.4. Finally, it is necessary to check that the limiting value of $\sigma_{ct} = f_{ctm}/\gamma_c$ is not exceeded. From Table 1.6, $f_{ctm} = 2.9$ N/mm^2 and thus $\sigma_{ct} = 2.9/1.5 = 1.93$ N/mm^2. From equation (5.7)

$$\tau_h = (1.25 \times 2/3 \times 1610 \times 400/1.15 - 18$$
$$\times 10^3)/400 \times 300 = 3.74 \text{ N/mm}^2$$

From equation (5.8)

$$\sigma_v = [(480 \times 10^3/2) + 503 \times 1.5$$
$$\times (400/1.15)]/400 \times 300 = 4.19 \text{ N/mm}^2$$

$$\sigma_h = 1133 \times (400/1.15)/400 \times 450 = 2.18 \text{ N/mm}^2$$

Thus σ_{ct} is given by

$$\sigma_{ct} = (-4.19 - 2.18)/2 \pm$$
$$\{[(-2.18 + 4.19)/2]^2 + 3.74^2\}^{1/2}$$
$$= -3.19 \pm 3.87 \text{ N/mm}^2$$

A positive sign denotes tension and thus $\sigma_{ct} = -3.19 + 3.87 = 0.68 < 1.93$ N/mm^2. Thus the upper limit of 1.93 N/mm^2 is not exceeded.

5.6 SUMMARY

Note that in the 1993 draft of EC8 the requirement that $\sigma_{ct(max)} \leqslant f_{ctm}/\gamma_c$ is satisfied if equations (5.11) and (5.12) are complied with. The principles of joint design set out in EC8 are in line with the procedures reported by Cheung *et al.* (1993) and are intended to minimize joint damage in a major earthquake. Damaged joints will cause a substantial reduction in the amount of energy that can be dissipated by the framing elements and they are virtually impossible to repair. The design of seismic joints necessitates stringent detailing requirements, and current guidelines may be obtained from ACI (1991) and Cheung *et al.* (1992) Typical details are given in Appendix H.

REFERENCES

ACI (1991) *Design of Beam–Column Joints for Seismic Resistance*, ACI SP-123.

Cheung P.C., Paulay T. and Park R. (1993) Behaviour of beam–column joints in seismically loaded reinforced concrete frames, *J. Inst. Struct. Eng.*, **71**, No. 8 (20 April)

NCE (1993) Finding fault, *New Civil Eng.*, 21 May.

SECTION ANALYSIS (3): COLUMNS AND WALLS

6.1 REINFORCED CONCRETE COLUMN DESIGN

The basic function of a column is to carry axial (vertical) loads in reinforced concrete building frames. In addition to the axial loads, however, the columns are required to sustain bending moments induced from the beams. These moments usually act in two orthogonal directions. Normally the moment in one of the two directions is significant and the column reinforcement can safely be obtained using the uniaxial analysis.

The method used in this book is the one adopted in the British Code BS 8110 where the equivalent uniaxial moments are obtained in each of the two orthogonal directions using the given bending moments. The design reinforcement steel area used is the largest of the two values obtained from the two uniaxial analyses. Some other sources of relevant information are listed in the 'references and bibliography' at the end of this chapter

The section analysis of columns is complex. It is necessary for the designer to make a realistic estimate of the overall column dimensions at the preliminary design stage; see Appendix B. At this stage, the reinforcement requirements are not known and the column dimensions will be governed by architectural and structural considerations, including cover and, in particular for seismic actions, more onerous detailing requirements for both longitudinal and hoop reinforcement. The initial estimate of column dimensions should be generous and linked to ease of construction and the strong column–weak beam philosophy.

In framed structures the internal columns normally carry large axial loads with relatively small moments, while the external columns carry moderate loads with larger moments. The sizing of the column can be achieved by assuming that the magnitude of the design axial load N_{sd} is given as shown below:

$$N_{sd} \not> 0.45 A_c f_{cd} \qquad \text{for low-ductility buildings}$$

$$N_{sd} \not> 0.40 A_c f_{cd} \qquad \text{for medium-ductility buildings}$$

$$N_{sd} \not> 0.35 A_c f_{cd} \qquad \text{for high-ductility buildings}$$

where $f_{cd} = f_{ck}/1.5$ is the design concrete strength, N_{sd} is the design axial load and A_c is the column cross-sectional area.

EC2 primarily deals with low-ductility structures; the magnitude of the axial load recommended in order that the column can have its optimum bending moment carrying capacity is equal to $N_{sd} = 0.4 A_c f_{cd}$, which is close to the low-ductility case above.

6.2 COLUMN DESIGN APPROACH

EC2 approaches the design of columns by first classifying the *whole structure* into 'non-sway' and 'sway' depending on its sensitivity to second-order effects, and then proceeds with the design of *individual* columns.

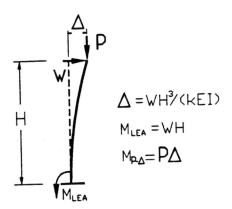

$$\Delta = WH^3/(kEI)$$

$$M_{\text{LEA}} = WH$$

$$M_{P\Delta} = P\Delta$$

Figure 6.1　Idealized non-sway structure.

6.2.1　Whole structures (non-sway)

(a)　Simplified approach

Structures can be deemed to be **non-sway** when design actions are not significantly affected by their deformation; the P–Δ effect is then deemed to be small. To look at this in its simplest form, consider the building to be idealized as a vertical cantilever of length H, flexural rigidity EI and subject to a horizontal load W and an axial thrust P (see Figure 6.1). The horizontal displacement, ignoring the effect of the axial load, is given by $\Delta = WH^3/kEI$. The calculated bending moment in this case is $M_{\text{LEA}} = WH$ (linear elastic analysis). The second-order P–Δ bending moment effect induced in the structure is given by $M_{P-\Delta} = P\Delta$. The P–Δ effects can be neglected when $M_{P-\Delta} \leqslant 0.1M_{\text{LEA}}$, which can be expressed using the above relationships in the form $H\sqrt{(P/EI)} \leqslant \sqrt{(0.1k)}$; for a pure cantilever $k = 3$ and hence the expression can be simplified to $H\sqrt{(P/\text{EI})} \leqslant 0.55$.

(b)　General approach

In normal cases, where the horizontal storey forces applied to the structure are small, the simplified approach normally satisfies the Code's requirement. If, however, large horizontal loads (seismic loads) are acting on the structure, a computer analysis may be necessary to establish the sway or non-sway status of the structure. To achieve this, it is first necessary to carry out the linear elastic analysis where the bending moments M_{LEA}, axial thrusts N_{sd} in the columns and the storey deflections Δ_{s} are recorded. A second analysis should be carried out with bending moments of magnitude $N_{\text{sd}} \Delta_{\text{s}}$ applied at the column joints and the bending moments in the columns $M_{P-\Delta}$ recorded. The EC2 condition

that $M_{P-\Delta} < 0.1M_{\text{LEA}}$ for non-sway status can be confirmed.

(c)　Structures with bracing elements

Based on the *simplified* approach, EC2 recommends that, when substantial bracing is provided in the form of walls, cores, or shear walls, the non-sway status of the structure requirement is satisfied, for buildings with n three or less storeys, as follows:

$$h_{\text{tot}}\sqrt{(F_{\text{v}}/E_{\text{cm}}I_{\text{c}})} \leqslant 0.2 + 0.1n$$

where h_{tot} is the total height of the structure in metres, F_{v} is the total axial service load (i.e. $\gamma_{\text{F}} = 1$) acting on the bracing elements, and $E_{\text{cm}} I_{\text{c}}$ is the sum of the flexural stiffnesses of all vertical bracing elements. For buildings with more than three storeys the condition becomes:

$$h_{\text{tot}}\sqrt{(F_{\text{v}}/E_{\text{cm}} I_{\text{c}})} \leqslant 0.6$$

Note that the above expression is very close to the one obtained above for a cantilever.

The concrete tensile stress of the bracing element should not exceed $f_{\text{ctk.0.05}}$. Bracing elements should be designed to carry 100% of the horizontal forces on the structure.

(d)　Structures without bracing elements

Framed structures with no substantial bracing elements could still be deemed to be of non-sway status when each column in a given storey that carries an average factored axial load of $N_{\text{a}} \nless (0.7 \gamma_{\text{F}} F_{\text{v}})/$(number of columns) has a slenderness ratio (effective length evaluation is based on non-sway status) *less than or equal to* the larger of the two values of λ given below:

$$\lambda = 25$$
$$\lambda = 15/\sqrt{(N_{\text{sd}}/A_{\text{c}}f_{\text{cd}})}$$

where λ is effective column height/radius of gyration and A_{c} is the column cross-sectional area.

6.2.2　Whole structures (sway)

Frames of approximately equal beam column stiffnesses, which do not satisfy the non-sway conditions but have an average storey slenderness ratio *smaller* than the larger of the following two values of λ, are deemed to be of **sway** status:

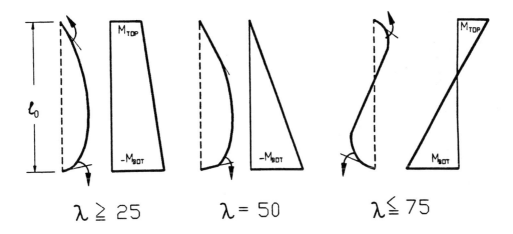

$$\lambda \geqq 25 \qquad \lambda = 50 \qquad \lambda \leqq 75$$

Figure 6.2   An isolated column in a non-sway frame: three different cases.

$\lambda = 50$
$\lambda = 25/\sqrt{(N_{sd}/A_c f_{cd})}$

6.2.3   Individual elements (non-sway and sway buildings)

Isolated columns in buildings, whether sway or non-sway, need not be designed for second-order effects if their slenderness ratio λ is *smaller* than the larger of the values of λ given below:

$\lambda = 25$
$\lambda = 15/\sqrt{(N_{sd}/A_c f_{cd})}$

6.2.4   Individual elements (non-sway buildings)

Second-order effects can be ignored in isolated columns in non-sway frames when their slenderness λ is within the value calculated by the expression $\lambda \leqslant 25(2 + M_{top}/M_{bot})$, where $|M_{top}| \leqslant |M_{bot}|$ (Figure 6.2); the signs of M_{top} and M_{bot} are positive if clockwise.

6.3   COLUMN SLENDERNESS RATIO

The slenderness ratio of a column depends on the way it is restrained at its ends and its cross-sectional dimensions. The **effective length** and the **radius of gyration** of the column need to be defined in this context.

6.3.1   Effective length

The effective length l_0 is dependent on the degree of restraint that beams and continuing columns provide at each end of the column length under consideration. Thus column/beam ratio coefficients K_{top} and K_{bot} at the top and bottom of the column are evaluated from the expression

$$K_{top}\ (K_{bot}) = \Sigma\ (I_c/l_c)/\Sigma\ (\alpha\ I_b/l_{b,eff})$$

where l_c is the length of the column measured between the centres of its restraint, $l_{b,eff}$ is the effective span of the beam, which takes into account the beam restraint at its other end by specifying the appropriate value of the coefficient α:

$\alpha = 1.0$ beam's other end fully fixed or an internal continuous support   A

$\alpha = 0.5$ beam's other end is a pin or a simple end support   B

$\alpha = 0$ beam is cantilever with its other end free   C

I_c is the second moment of area of the column section, and I_b is the second moment of area of the beam section.

The effective length l_0 in terms of l_c can be obtained from the relationship $l_0 = l_c\beta$ where the value of β is obtained for the non-sway and sway frames respectively from Figure 6.3, using the relevant values of K_{top} and K_{bot}.

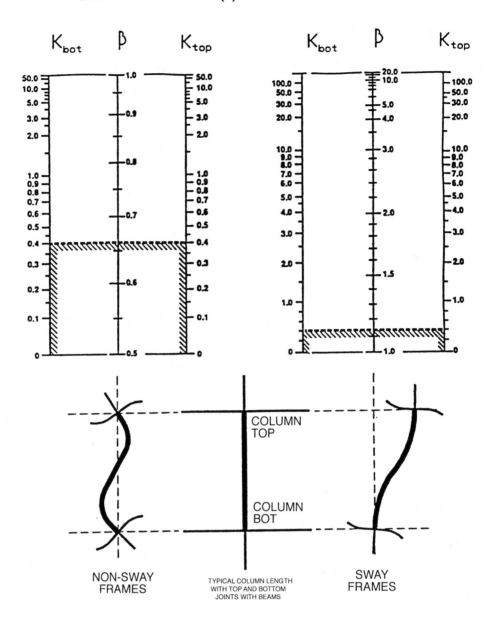

Figure 6.3 Charts showing β values for non-sway and sway frames.

6.3.2 Radius of gyration

The **radius of gyration** $r = \surd(I_c/A_c)$ where A_c is the cross-sectional area of the column. For rectangular columns of width b and depth h, where h and b are the major and minor axes respectively, the radii of gyration are $r_{major} = h/\surd 12$ and $r_{minor} = b/\surd 12$. For circular columns of diameter D, the radius of gyration is $r_{cir} = D/4$.

6.3.3 Slenderness ratio

The **slenderness ratio** of a column is given by $\lambda = (l_0/r)$ where l_0 and r are the effective length and radius of gyration defined above. Slenderness ratios for rectangular and circular columns are as follows:

$\lambda_{major} = 3.464\ l_0/h$ rectangular column about the major axis

$\lambda_{cir} = 4l_0/D$ circular column

6.4 COLUMN DESIGN ACTIONS

The actions that are normally involved in the design of a column are as follows.

6.4.1 Linear elastic analysis results

Bending moments at the two ends of the column, and the corresponding axial thrust, are obtained from a linear elastic analysis for the worst load combinations.

6.4.2 Structure's imperfections allowance

Additional bending moments to be added to those obtained from the analysis are allowed for, to cater for dimensional inaccuracies in the position and line of action of the axial loads in the structure (Figure 6.4). Owing to imperfections in the construction, the structure should be designed to carry additional horizontal loads

$$F_{Hi} = \theta \Sigma V$$

where

$$\Sigma (V) = [1/(100 \sqrt{h_{tot}})]\alpha$$

$\Sigma(V)$ is the total vertical load above the ith floor; $\alpha = 1$ when bracing elements exist, otherwise $\alpha = \sqrt{[1+1/(\text{number of columns})]/2}$ when only columns are present; $\theta < 0.0025$ when second-order effects are ignored, and $\theta \nleq 0.005$ when second-order effects are considered.

(a) Non-bracing members

The bending moment due to imperfections is

$$M_{IMP} = \theta N_{sd} l_0/2$$

(b) Bracing members

The imperfections' effect on bracing members is normally small but can nevertheless be taken into account in the linear elastic analysis by introducing in the analysis additional horizontal storey forces F_{HM} given by

$$F_{HM} = (N_{sd \text{ (above floor)}} + N_{sd \text{ (below floor)}}) \theta/2$$

6.4.3 Second-order effect

In sway frames, columns of slenderness ratio within the values specified in sections 6.2.2 and 6.2.3, and in non-sway frames, columns of slenderness ratio above the value specified in section 6.2.4, should be designed by adding the second-order effect.

The P–Δ moment effect M_s is evaluated as follows:

$$M_s = N_{sd} \Delta$$

where

$$\Delta = k_1 (l_0)^2/(10R)$$

Here Δ is the deflection causing the P–Δ effect, $1/R$ is the radius of curvature at the section of deflection Δ and

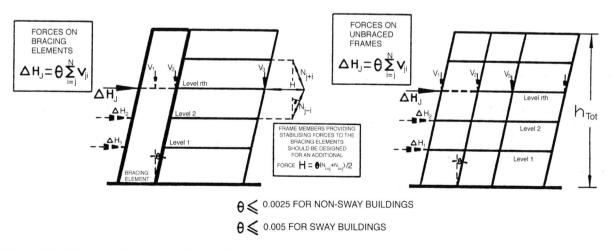

Figure 6.4 Allowance for structural imperfections.

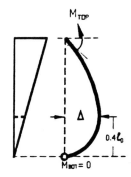

DESIGN MOMENT = $0.4 M_{TOP} + P\Delta$

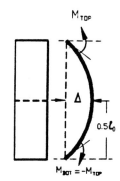

DESIGN MOMENT = $M_{TOP} + P\Delta$

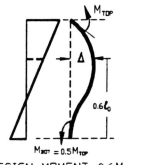

DESIGN MOMENT = $0.6 M_{TOP} - 0.4 M_{BOT} + P\Delta$

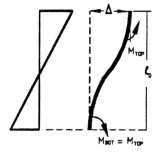

DESIGN MOMENT = $M_{TOP} + P\Delta$

Figure 6.5 Column design moment in four different cases.

$$k_1 = (\lambda/20) - 0.75 \quad \text{for } 15 \le \lambda \le 35$$

$$k_1 = 1 \qquad\qquad \text{for } \lambda > 35$$

Note: d_0 and λ are the corresponding governing effective depth and slenderness ratios in either the minor or major direction. R is given by

$$1/R = 2k_2 \, (f_{yk}/1.15)/(0.9 E_s \, d_0)$$

with

$$k_2 = 1$$

This is a conservative value; a more accurate value is given below:

$$k_2 = [(0.567 \, A_c \, f_{ck} + A_s \, f_{yk}/1.15) - N_{sd}]/$$
$$[(0.567 \, A_c \, f_{ck} + A_s \, f_{yk}/1.15) - 0.4 \, A_c f_{ck}/1.5]$$

6.4.4 Column design moment

Columns for which second-order effects can be neglected are always designed for the largest moment from the linear elastic analysis plus the moment from the imperfections' calculation. When,

however, the second-order effects are significant, the column should be designed for the maximum moment at the section at which the P–Δ effect is greatest, as illustrated by Figure 6.5

Assume that the largest absolute bending moment at the top of the column from the linear elastic analysis plus that due to the imperfections of the building is represented by M_{top} (M_{top} is interchanged with M_{bot} in the formulae below if the absolute value of the latter is the larger of the two values).

The design moment is the corresponding value from the expressions given below, noting that $P\Delta = M_s$:

$$0.4 \, M_{top} + P\Delta \qquad\qquad \text{or } M_{top} \text{ if larger}$$

$$M_{top} + P\Delta$$

$$0.6 \, M_{top} - 0.4 \, M_{bot} + P\Delta \qquad \text{or } M_{top} \text{ if larger}$$

6.5 DESIGN STEPS

The flow diagrams shown in Figures 6.6 and 6.7 trace the design steps that must be followed in the design of columns. Figure 6.6 shows that for a whole structure, whereas Figure 6.7 is for isolated members.

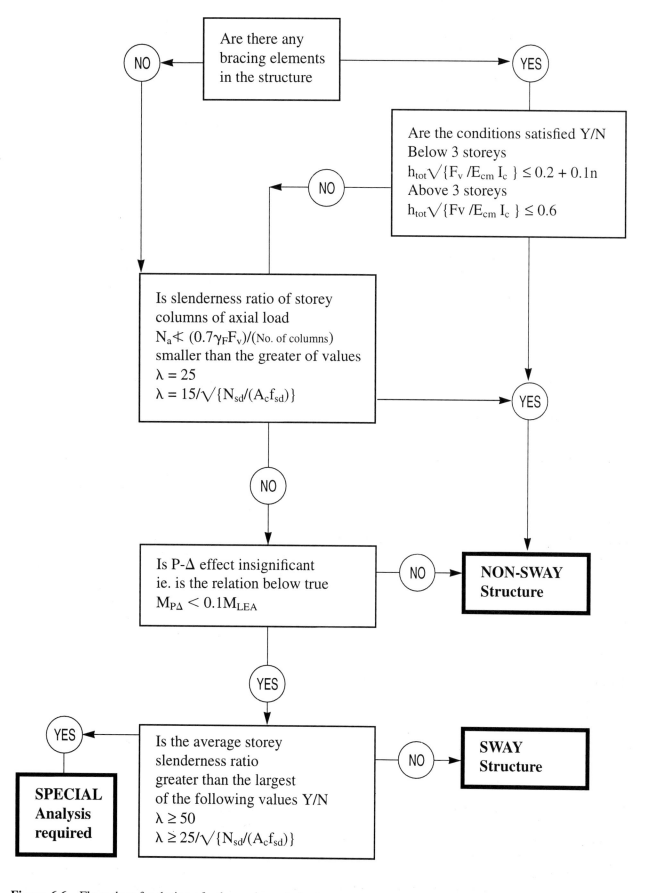

Figure 6.6 Flow chart for design of columns in a whole structure.

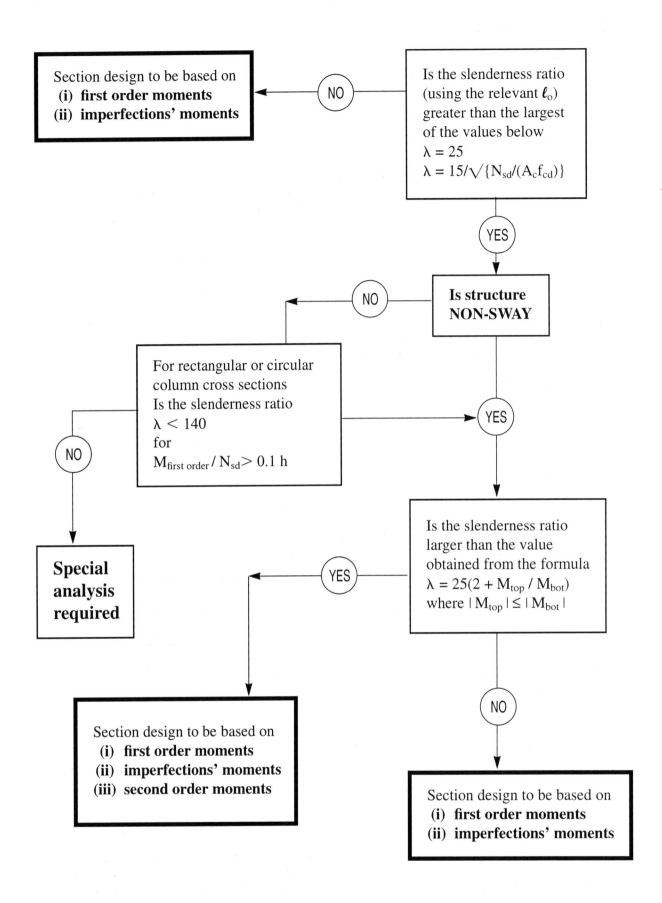

Figure 6.7 Flow chart for design of columns or isolated members.

Example 6.1: design example

A six-storey frame has a column grid of 7 m × 5 m (similar to that shown later in Figure 8.1), noting that in this example the column heights are 3 m with the exception of the ground to first floor, which is 5 m. In the transverse X direction, the building is assumed to have no bracing elements (shear or core walls). Given the following data, check whether the frame is sway or non-sway:

$N_{sd} = 1260$ kN   (exterior column)
$N_{sd} = 2520$ kN   (interior column)
$b = 350$ mm   (column width)
$I_c/I_b = 1$ (ratio of column to beam second moments of area)

$f_{ck} = 30$ N/mm²

The columns are subjected to axial thrust and bending and, as a starting point, the exterior column is proportioned with a view to sustaining an optimum moment.
Using $N_{sd} = 0.4 A_c f_{cd}$, then

$A_c = N_{sd}/(0.4f_{ck}/1.5) = 157\,500$ mm²
$h = 157\,500/350 = 450$ mm

Structure has no bracing elements

The factor

$(0.7\gamma_F F_v)/(\text{number of columns})$

$= 0.7 (\gamma_F F_v)/4 = 0.175(\gamma_F F_v)$

and

load on an exterior column

$= (\gamma_F F_v)/6 = 0.167 (\gamma_F F_v)$

load on an interior column

$= (\gamma_F F_v)/3 = 0.333 (\gamma_F F_v)$

Slenderness ratio of internal columns should be checked and compared to

$\lambda_1 = 25$
$\lambda_2 = 15/\sqrt{(N_{sd}/A_c f_{cd})}$

Ground to first floor

We have for the interior column

$K_{bot} = 0.4$ (minimum permissible)

$K_{top} = [(1/3 + 1/5)/(2/7)] (I_c/I_b) = 1.87$

$\beta = 0.73$

The effective column length

$l_0 = \beta l_c = 0.73 \times 5000 = 3650$ mm

Radius of gyration in major direction

$r_{major} = h/\sqrt{12} = 450/\sqrt{12} = 129.9$ mm

Slenderness ratio

$\lambda_{major} = 3650/129.9 = 28.1$

$\lambda_2 = 15/\sqrt{[2520\,000/(350 \times 450 \times 30/1.5)]} = 16.77$

$\lambda_1 > \lambda_2$ and $\lambda > \lambda_2$. Structure is a sway frame.

First to second floor

Assuming the beam and column stiffnesses are based on rectangular sections

$K_{bot} = 1.87$

$K_{top} = [(1/3 + 1/3)/(2/7)] (I_c/I_b) = 2.33$

$\beta = 0.85$

The effective column length

$l_0 = \beta l_c = 0.85 \times 3000 = 2550$ mm

Radius of gyration in major direction

$r_{major} = h/\sqrt{12} = 450/\sqrt{12} = 129.9$ mm

Slenderness ratio

$\lambda_{major} = 2550/129.9 = 19.6$

$\lambda_1 > \lambda_2$ and $\lambda < \lambda_1$. Structure is a non-sway frame.

Adjustment to cross-sectional dimensions

It advisable at this stage to enlarge the ground to first floor column by a small amount to make the frame a non-sway one. Assuming new dimensions to be $b = 350$ mm and $h = 550$ mm and rechecking the λ values:

$K_{\text{bot}} = 0.4$ (minimum permissible)

$K_{\text{top}} = [(1/3 + 1/5)/(2/7)] \, (1.83 I_c/I_b) = 3.42$

$\beta = 0.77$

The effective column length

$l_0 = \beta \, l_c = 0.77 \times 5000 = 3850 \, \text{mm}$

Radius of gyration in major direction

$r_{\text{major}} = h/\sqrt{12} = 550/\sqrt{12} = 158.77 \, \text{mm}$

Slenderness ratio

$\lambda_{\text{major}} = 3850/158.77 = 24.25$

$\lambda_2 = 15/\sqrt{[2520\,000/(350 \times 550 \times 30/1.5)]}$

$= 18.54$

$\lambda_1 > \lambda_2$ and $\lambda < \lambda_1$. Structure is a non-sway frame.

6.6 COLUMN DESIGN SECTION

6.6.1 Unconfined concrete stress–strain diagram

The stress–strain diagram is shown in Figure 6.8, and concrete strengths (N/mm²) are shown in Table 6.1, where f_{ck} is cylinder characteristic strength, f_{ctm} is mean concrete tensile strength, $f_{\text{ctk,0.05}}$ is minimum

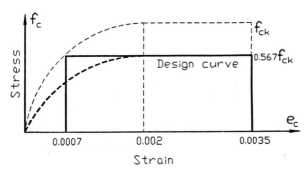

Figure 6.8 Stress–strain diagram for case of unconfined concrete.

Table 6.1 Concrete strengths (N/mm²) in case of unconfined concrete

f_{ck}	16.0	20.0	25.0	30.0	35.0	40.0	45.0	50.0
f_{ctm}	1.9	2.2	2.6	2.9	3.2	3.5	3.8	4.1
$f_{\text{ctk,0.05}}$	1.3	1.5	1.8	2.0	2.2	2.5	2.7	2.9
f_{cu}	20.0	25.0	30.0	37.0	45.0	50.0	55.0	60.0

concrete tensile strength and f_{cu} is cube characteristic strength.

6.6.2 Steel reinforcement design stress–strain diagram

In this case, the stress–strain diagram is shown in Figure 6.9. The steel strengths (N/mm²) are shown in Table 6.2, where f_{yk} is characteristic tensile strength of reinforcement and f_{yck} is characteristic compressive strength of reinforcement. The steel strains (% minima) are shown in Table 6.3, where e_{su} is per cent ultimate reinforcement steel strain and DL is ductility level of section.

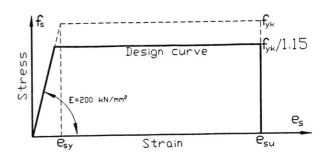

Figure 6.9 Stress–strain diagram for case of a design with steel reinforcement.

Table 6.2 Steel strengths (N/mm²) in design with reinforcement

f_{yk}	220	250	400	460	500
f_{yck}	220	250	400	400	400

Table 6.3 Steel strains (% minima) in design with reinforcement

e_{su}	6%	9%	12%
DL	I (low)	II (medium)	III (high)

6.6.3 Rectangular reinforced concrete column section analysis

The cross-section and other diagrams for this case are shown in Figure 6.10. The analysis of the section subject to an axial load N and a uniaxial bending moment M is carried out assuming the following parameters are given: h = overall depth of the section, b = overall breadth of the section, x_{si} = reinforcement depth factor, f_{ck} = concrete characteristic strength, f_{yk} = tensile characteristic steel strength and f_{yck} = compressive characteristic steel strength.

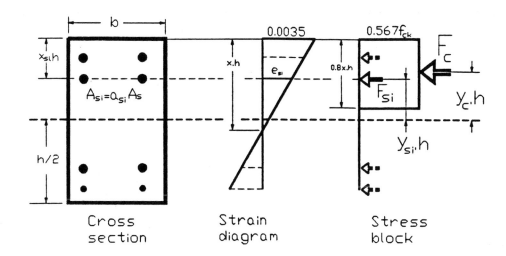

Figure 6.10 Analysis of rectangular reinforced concrete column section.

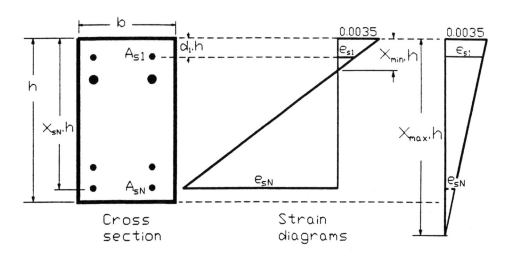

Figure 6.11 Neutral axis depth factor of rectangular reinforced concrete column section.

Neutral axis depth factor

Consider the diagrams in Figure 6.11. The minimum value of the neutral axis depth factor ensures that the stress f_{s1} in the compression steel does not fall below $(f_{yck}/1.15)$. This is given by the expression

$$x_{min} = x_{s1}/(1 - f_{yck}/805)$$

As x increases, the value of f_{sN} decreases until it reaches zero; further increases in x cause f_{sN} to change sign and become compressive in nature. The maximum value of the neutral axis depth factor x is the lesser of the values below and ensures that:

- if f_{sN} is greater than or equal to $-f_{yck}/1.15$, then $x_{max} = x_{sN}/(1 - f_{yck}/805)$;
- $x_{max} \leqslant 1.25$ ensures that the concrete compression block lies within the section.

Steel reinforcement stresses

The steel stress f_{si} of the ith bar located at a depth of x_{si} for a given value of x is given by $f_{si} = 700[1 -(x_{si}/x)]$. If the strain e_{si} of a steel bar located at the distance x_{si} is negative for a given value of x and f_{si} becomes less than $-f_{yk}/1.15$ then $f_{st}>$ is taken as $-f_{yk}/1.15$.

Limits of compressive stress f_{s1}

The compressive stress f_{s1} is subject to the neutral axis depth factor. From the strain diagram of the column section (Figure 6.10) $e_{s1} = 0.0035 [1 -(x_{s1}/x)]$. The steel stress $f_{s1} = Ee_{s1}$; f_{s1} is set equal to $f_{yck}/1.15$ and from this the expression for x_{min} above is obtained.

Limits of compressive stress f_{sN}

The tensile strain e_{sN} of the steel bar at the distance

x_{sN} decreases as the neutral axis depth factor x increases; if the value of $f_{sN} = 700[1-(x_{sN}/x)]$ becomes less than $f_{yck}/1.15$ then f_{sN} is taken as $f_{yck}/1.15$.

(a) Equilibrium equations

Vertical equilibrium
We have

$$\rho = 100A_s/A_c$$

where A_s and A_c are the total areas of steel reinforcement and concrete cross-section respectively. Also

$$N/A_c = F_c/A_c + \rho(\Sigma f_{si}a_{si})/100$$

$$F_c/A_c = 0.453xf_{ck}$$

Moment equilibrium
We have

$$M/(hA_c) = M_c/(hA_c) + \rho(\Sigma f_{si}a_{si}y_{si})/100$$

$$M_c/(hA_c) = (F_c/A_c)(0.5 - 0.4x)$$

where N and M are respectively the design axial thrust and bending moment acting on the column section. Since the units of f_{ck}, f_{yk} and f_{yck} are normally N/mm², the units of A_c, h and y_{si} are chosen to be in mm units and the axial load and moment in kN and kN mm units.

(b) Charts of v against μ for various values of ρ

The percentage steel reinforcement for a given set of values of the axial force and bending moment M is obtained by using the charts of the plot of $v = N/A_c$ against $\mu = M/(hA_c)$ (both in N/mm²) for various values of ρ, specified positions of the steel and given f_{ck}, f_{yk} and f_{yck}.

Charts 1 and 2 are given in Appendix F for $f_{yk} = 400$, $f_{yck} = 400$ and $f_{yk} = 460$, $f_{yck} = 400$ for values of $x_{s1} = 0.1$, $x_{sN} = 0.9$ and $a_{s1} = a_{sN} = 0.5$.

(c) Percentage of steel evaluation

See computer program 1.

6.6.4 Circular reinforced concrete column section analysis

Diagrams relating to this case are shown in Figure 6.12. The limits for the neutral axis depth factor x and limits on the steel stresses in the reinforcement are the same as for the rectangular column cross-section.

(a) Equilibrium equations

Vertical equilibrium
We have

$$\rho = 100 \, A_s/A_c$$

$$N/A_c = F_c/A_c + \rho \, (\Sigma f_{si}a_{si})/100$$

Moment equilibrium
We have

$$M/(dA_c) = M_c/(dA_c) + \rho \, (\Sigma f_{si}a_{si} y_{si})/100$$

and F_c/A_c is evaluated as follows:

$$k_1 = 0.5 - 0.8x$$

$$k_2 = \sqrt{(0.25 - k_1{}^2)}$$

$$\theta = \tan^{-1}(k_1/k_2)$$

$$F_c/A_c = (4/\pi)(0.25 - k_1 k_2)(0.567 f_{ck})$$

Also

$$M_c/(dA_c) = 4/(3\pi)(0.25 \sin\theta - 2k_2 k_1{}^2)(0.567 f_{ck})$$

where n is the number of bars in the cross-section, A_s is the total steel area placed uniformly at a diameter of $(1-d_1)d$, A_s/n is the cross-sectional area of one bar and $\varphi = 2\pi/n$ is the angle between reinforcing bars.

Finally

$$y_{si} = 0.5(1 - d_1) \cos[(i-1)\varphi]$$

$$x_{si} = 0.5 - y_{si}$$

$$f_{si} = 700[1 - (x_{si}/x)]$$

$$F_s/A_c = \rho \, (\Sigma f_{si})/(100n)$$

$$M_s/(dA_c) = \rho \, (\Sigma f_{si}y_{si})/(100n)$$

```
INPUT "Design bending moment in kN.m = "; DM
INPUT "Design axial force in kN = "; DN
INPUT "Overall section width in mm = "; DB
INPUT "Overall section depth in mm = "; DD


FYK = 460: FCK = 30: FYCK = 400
NU = DN * 1000/(DB * DD): MU = DM * 1000000/(DB * DD * DD)
XS1 = 0.1: XSN = 1 - XS1: YS1 = 0.5 - XS1: YSN = 0.5 - XSN
K = 0.8: Q = 0.85/1.5: N = 2

SCREEN 12
VIEW (1, 1)-(638, 450), 1, 2
WINDOW (0, 0)-(16, 50)

DIM ASI(N)
XMIN = XS1: xmax = XSN/(1 - FYCK/805)
xsi(1) = XS1: ASI(1) = 0.5
xsi(2) = XSN: ASI(2) = 0.5

FOR i = 1 TO N: aos = aos + ASI(i): NEXT
FOR i = 1 TO N: ASI(i) = ASI(i)/aos: NEXT

FOR ro = 0 TO 8 STEP 1
FOR X = XMIN TO xmax STEP 0.025
fs = 0: ms = 0
        FOR i = 1 TO N
        xsi = xsi(i): ASI = ASI(i): ysi = 0.5 - xsi: fsi = 700 * (1 - xsi/X)
        IF fsi >= 0 AND fsi > FYCK/1.15 THEN fsi = FYCK/1.15
        IF fsi < 0 AND fsi < -FYK/1.15 THEN fsi = -FYK/1.15
        fs = fs + fsi * ASI: ms = ms + fsi * ysi * ASI
        NEXT
    XO = X: IF X > 1.25 THEN XO = 1.25
    fc = Q * K * XO * FCK
    mc = Q * K * XO * FCK * (0.5 - 0.5 * K * XO)
    yn = fc + ro * fs/100
    xm = mc + ro * ms/100
    IF X = XMIN THEN
    PSET (xm, yn)
    ELSE
    LINE -(xm, yn)
    END IF
    NEXT
NEXT

FOR i = 0 TO 16 STEP 1
PSET (i, 0): LINE -(i, 50), 2
NEXT

FOR j = 0 TO 50 STEP 5
LINE (0, j)-(16, j), 2
NEXT

LINE (0, 0.1 * FCK/1.5)-(16, 0.1 * FCK/1.5)
LINE (0, NU)-(16, NU),2: LINE (MU, 0)-(MU, 50), 2
END
```

Computer program 1

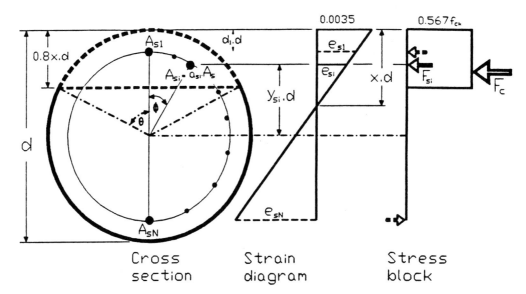

Figure 6.12 Analysis of circular reinforced concrete column section.

(b) Charts of v against μ for various values of ρ

The percentage steel reinforcement for a given set of values of the axial force N and bending moment M is obtained by using the charts of the plot of $v = N/A_c$ against $\mu = M/(dA_c)$ (both in N/mm²) for various values of the percentage steel reinforcement ρ, specified positions of the steel and given f_{ck}, f_{yk} and f_{yck}.

Charts 3 and 4 are given in Appendix F for $f_{yk} = 400$ $f_{yck} = 400$ and $f_{yk} = 460$, $f_{yck} = 400$ for values of $d_{s1} = 0.1$.

A complete set of software can be obtained from the authors.

(c) Percentage of steel evaluation

See computer program 2.

6.6.5 Biaxial column section analysis

The majority of column sections are normally subject to bending moments in two orthogonal directions and the reinforcement should be derived from a biaxial analysis. The method employed here is that currently used in the British Code, where the moments in the two orthogonal axes of the column are replaced with an equivalent moment in one of the axes subject to certain constraints stated below.

When it is necessary to consider biaxial bending in rectangular column sections (Figure 6.13), the design charts derived above for rectangular sections can be employed provided the equivalent uniaxial

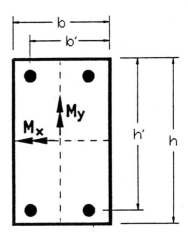

Figure 6.13 Analysis of rectangular reinforced column under biaxial bending.

moment M'_x (or M'_y) obtained from the expressions below is used:

$$M'_x = M_x + \beta_N (h'/b')M_y \quad \text{for } M_x/M_y \geqslant h'/b'$$

$$M'_y = M_y + \beta_N (b'/h')M_x \quad \text{for } M_x/M_y < h'/b'$$

where β is given in Table 6.4 as a function of axial ratio $N/(bhf_{ck})$.

Table 6.4 Value of β as a function of axial ratio

$N/(bhf_{ck})$	0.000	0.125	0.250	0.375	0.500	0.625	$\geqslant$0.750
β_N	1.000	0.880	0.770	0.650	0.530	0.420	0.300

```
INPUT "Design bending moment in kN.m = "; DM
INPUT "Design axial force in kN = "; DN
INPUT "Overall section diameter in mm = "; DD

fck = 30: fyk = 400
IF fyk > 400 THEN fyck = 400 ELSE fyck = fyk
AC = 3.1414 * DD * DD/4
NU = 1000 * DN/AC
MU = 1000000 * DM/(DD * AC)

d1 = 0.1: n = 8: DIM ysi(n)
xmin = d1/(1 - fyck/805): xmax = 1.25
p = (fyck/805 - d1od)/(1 - fyck/805): IF p < 0.25 THEN xmax = 1 + p
FOR i = 1 TO n
ysi(i) = 0.5 * (1 - d1) * COS(((i - 1) * (2 * 3.1415/n)))
NEXT

SCREEN 12
VIEW (1, 1)-(600, 400), 1, 2
WINDOW (0, 0)-(12, 50)

FOR ro = 0 TO 8 STEP 1
FOR x = xmin TO xmax STEP 0.025
fs = 0: ms = 0
            FOR i = 1 TO n
            ysi = ysi(i): xis = 0.5 - ysi
            fsi = 700 * (1 - xis/x)
            IF fsi >= 0 AND fsi >= fyck/1.15 THEN fsi = fyck/1.15
            IF fsi < 0 AND fsi < -fyk/1.15 THEN fsi = -fyk/1.15
            fs = fs + fsi
            ms = ms + fsi * ysi
            NEXT
    k1 = 0.5 - 0.8 * x
    k2 = SQR(0.25 - k1 * k1)
    theta = ATN(k2/k1): IF k1 < 0 THEN theta = theta + 3.1415
    fc = (0.25 * theta - k1 * k2) * (0.567 * fck)
    mc = (0.567 * fck) * ((0.25 * SIN(theta) - 2 * k1 * k2 * k1))/3
    yn = fc + (3.1415/(400 * n)) * fs * ro
    xm = mc + (3.1415/(400 * n)) * ms * ro
    yn = yn * 4/3.1415: xm = xm * 4/3.1415
    IF x = xmin THEN PSET (xm, yn) ELSE LINE -(xm, yn)
    NEXT
NEXT

FOR i = 0 TO 12 STEP 1
PSET (i, 0): LINE -(i, 50), 2
NEXT
FOR j = 0 TO 50 STEP 5
PSET (0, j): LINE -(12, j), 2
NEXT
PSET (0, NU): LINE -(12, NU), 4
PSET (MU, 0): LINE -(MU, 50), 4
END
```

Computer program 2

Example 6.2: rectangular column section, with no sway

In this example, we shall consider a column with frame non-sway in both x and y directions. The dimensional details are:

$b = 250$ $b' = 210$

$h = 400$ $h' = 360$

column length $l_c = 3000$

$\beta_x = 0.75$ $\beta_y = 0.85$

Bending moments including imperfections' effect (kN m) and axial load (kN):

moments at bottom of column

$M_x = 180$ $M_y = 80$

moments at top of column

$M_x = 36$ $M_y = 40$

Axial load = 1125. The column is shown in Figure 6.14.

Significance of second-order effect

We have

$\lambda_x = 0.75 \times 3000 \times 3.46/400 = 19.49$

$\lambda_y = 0.85 \times 3000 \times 3.46/250 = 35.33$

axial ratio

$= N/(bhf_{ck}) = 1125\,000/(250 \times 400 \times 30)$

$= 0.375$

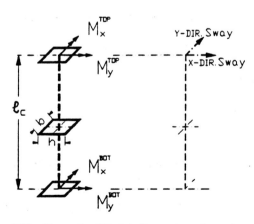

Figure 6.14 Examples 6.2–6.4: the rectangular column.

slenderness ratio factor

$= 15/\sqrt{[N/(bhf_{ck}/1.5)]}$

$= 15/\sqrt{(0.375 \times 1.5)} = 20$

$\lambda_x < 25$ column short

$\lambda_y > 25$ column slender

$\lambda_{crit} = 25(2 + 50/100) = 62.5$

$\lambda_y < 62.5$

Second-order effect can be ignored.

Section analysis

We have

$h'/b' = 360/210 = 1.71$

$M_x/M_y = 180/80 = 2.25$

Therefore

$M_x/M_y > h'/b'$

$\beta_N = 0.65$ corresponds to an axial ratio = 0.375, from Table 6.4. Thus

$M = M_x + \beta_N M_y \, (h'/b')$

$= 180 + 0.65 \times 80 \times 1.71 = 268.9$

$M/bh^2 = 268.9/(0.4^2 \times 250)$

$= 6.72 \text{ N/mm}^2$

$N/bh = 1125/(0.4 \times 250)$

$= 11.25 \text{ N/mm}^2$

From chart 2, Appendix F

$\%\rho = 3.9$

$A_s = 0.039 \times 400 \times 250 = 3900 \text{ mm}^2$

Provide four 25 mm diameter bars $f_{yk} = 400 \text{ N/mm}^2$ at each short end side of the section, at an embedment of 40 mm.

Example 6.3: rectangular column section, with no sway, longer column

In this example, we consider a similar case to that in Example 6.2, but for a longer column, so that we see the significance of second-order effects. The dimensional details are:

$$b = 250 \quad b' = 210$$

$$h = 400 \quad h' = 360$$

column length $\quad l_c = 5500$

$$\beta_x = 0.75 \quad \beta_y = 0.85$$

Bending moments including imperfections' effect (kN m) and axial load (kN):

moments at bottom of column

$$M_x = 180 \quad M_y = 80$$

moments at top of column

$$M_x = 36 \quad M_y = 40$$

Axial load = 1125.

Significance of second-order effect

We have

$$\lambda_x = 0.75 \times 5500 \times 3.46/400 = 35.73$$

$$\lambda_y = 0.85 \times 5500 \times 3.46/250 = 64.78$$

axial ratio

$$= N/(bhf_{ck}) = 1125\,000/(250 \times 400 \times 30)$$
$$= 0.375$$

slenderness ratio factor

$$= 15/\sqrt{[N/(bhf_{ck}/1.5)]}$$
$$= 15/\sqrt{(0.375 \times 1.5)} = 20$$

$$\lambda_x > \quad 25 \text{ column slender}$$

$$\lambda_{crit} = 25(2 + 36/180) = 55$$

$$\lambda_x < 55$$

Second-order effect can be ignored in this direction.

$$\lambda_y > \quad 25 \text{ column slender}$$

$$\lambda_{crit} = 25(2 + 40/80) = 62.5$$

$$\lambda_y > 62.5$$

Second-order effect should be accounted for in this direction.

Section analysis

The design moments are

$$M_x = 180$$

$$M_y = (0.6 \times 80 - 0.4 \times 40) + M_s$$
$$= 32 + M_s \quad \text{(see Figure 6.5)}$$

where $M_s = N\Delta$. Young's modulus

$$E = 9.5(f_{ck} + 8)^{1/3} = 9.5 \times (38)^{1/3}$$
$$= 31940 \text{ N/mm}^2$$

$$\Delta = [(0.85 \times 5500)^2/10](1/R)$$
$$= 2.186 \times 10^6(1/R)$$

$$1/R = 2 \times (400/1.15)/(0.9 \times 31940 \times 360)$$
$$= 6.722 \times 10^{-5}$$

$$\Delta = 21.86 \times 6.722 = 147$$

$$M_s = 1125 \times 0.147 = 165.4 \text{ kN m}$$

$$M_y = 197.4 \text{ kN m}$$

$$h'/b' = 360/210 = 1.71$$

$$M_x/M_y = 180/197.4 = 0.91$$

Therefore

$$M_x/M_y < h'/b'$$

This corresponds to an axial ratio = 0.375, from Table 6.4. Thus

$$M = M_y + \beta_N M_y (b'/h')$$
$$= 197.4 + 0.65 \times 180/1.71 = 265.8$$

$$M/bh^2 = 265.8/(400 \times 0.25^2) = 10.63 \text{ N/mm}^2$$

$$N/bh = 1125/(0.4 \times 250) = 11.25 \text{ N/mm}^2$$

From chart 2

$$\%\rho = 5.6 \quad A_s = 0.056 \times 400 \times 250$$
$$= 5600 \text{ mm}^2$$

Re-evaluation of the factor k_2, which was taken as equal to unity, is now necessary:

$$N_u = 0.57 \times 30 \times 400 \times 250 + 5600 \times 400/1.15$$
$$= 3658 \text{ kN}$$

$N_u - N = 3658 - 1125 = 2533$ kN

$N_{bal} = 0.4 \times (30/1.5) \times 400 \times 250 = 800$ kN

$N_u - N_{bal} = 2858$

$k_2 = 2533/2858 = 0.886$

$M_s = 0.886 \times 165.4 = 146.5$ kN m

$M_y = 178.4$

$M = 246.8$

$M/bh^2 = 9.87$

$\%\rho = 5.3$

$A_s = 5300$ mm^2

Rechecking the value of k_2:

$N_u = 3554$

$k_2 = 2429/2754 = 0.88$

This is close to above value; therefore the steel area is $A_s = 5300$ mm^2. Provide six bars 25 mm diameter on each of the long edges of the column section at 40 mm embedment.

It should be noted that the reinforcement required is rather high and the sensible solution is to increase the width of the section to ensure that the second-order effect in the minor direction can be ignored and hence design for the major direction equivalent moments.

Assume $b = 275$ mm, $b' = 235$ mm and effective length factor is taken as $\beta_y = 0.85$; $\lambda_y = 0.85 \times 5500 \times 3.46/275 = 58.9$ and hence $\lambda_y < \lambda_{crit}$; this makes it possible to ignore second-order effects in the minor direction; the axial ratio $N/(bhf_{ck}) = 0.341$ and consequently from Table 6.4 $\beta_N = 0.68$. Also, $M_x/M_y > h'/b'$; hence the design moment $M = 180 + 0.68 \times 80 \times (360/235) = 263$ kN m and $M/(bh^2) = 5.36$. Thus, $\%_\rho = 3.0$ and $A_s = 0.03 \times 400 \times 275 = 3300$ mm^2. Provide four bars 25 mm diameter on each of the short edges of the column section at 40 mm embedment, which gives a more satisfactory solution.

Example 6.4: rectangular column section, with sway in one direction

This is similar to Example 6.2, but with frame non-sway in the x direction and sway in the y direction. The dimension details are:

$b = 250$ $b' = 210$

$h = 400$ $h' = 360$

column length $l_c = 3000$

$\beta_x = 1.5$ $\beta_y = 0.85$

Bending moments including imperfections' effect (kN m) and axial load (kN):

moments at bottom of column

$M_x = 180$ $M_y = 80$

moments at top of column

$M_x = 36$ $M_y = 40$

Axial load $= 1125$.

Significance of second-order effect

We have

$\lambda_x = 1.5 \times 3000 \times 3.46/400 = 38.97$

$\lambda_y = 0.85 \times 3000 \times 3.46/250 = 35.33$

axial ratio

$= N/(bhf_{ck}) = 1125\,000/(250 \times 400 \times 30)$

$= 0.375$

slenderness ratio factor

$= 15/\sqrt{[N/(bhf_{ck}/1.5)]} = 15/\sqrt{(0.375 \times 1.5)}$

$= 20$

$\lambda_x > 25$

$\lambda_x < 140$

$M_{first\ order}/(N \times h) > 0.1$

Second-order effect should be taken into account in the x direction.

$\lambda_y > 25$ column slender

$\lambda_{crit} = 25(2 + 40/80) = 62.5$

$\lambda_y < 62.5$

Second-order effect can be ignored in the y direction.

Section analysis

Design moments are

$M_x = 180$

$M_x = 180 + M_s$ (see Figure 6.5)

where $M_s = N\Delta$. Young's modulus

$E = 31940$ N/mm²

$\Delta = [(0.85 \times 3000)^2/10](1/R) = 0.65 \times 10^6(1/R)$

$1/R = 2 \times (400/1.15)/(0.9 \times 31940 \times 360)$

$\quad = 6.722 \times 10^{-5}$

$\Delta = 43.7$

$M_s = 1125 \times 0.0437 = 49.15$ kN m

$M_x = 229.15$ kN m

$h'/b' = 360/210 = 1.71$

$M_x/M_y = 229.15/80 = 2.86$

Therefore

$M_x/M_y > h'/b'$

This corresponds to an axial ratio = 0.375, from Table 6.4, $\beta_N = 0.65$. Thus

$M = M_x + \beta_N M_y \, (b'/h')$

$\quad = 229.15 + 0.65 \times 80 \times 1.71 = 318$

$M/bh^2 = 318/(0.4^2 \times 250) = 7.95$ N/mm²

$N/bh = 1125/(0.4 \times 250) = 11.25$ N/mm²

From chart 2

$\%\rho = 4.7$

$A_s = 0.047 \times 400 \times 250 = 4700$ mm²

Re-evaluation of the factor k_2, which was taken as equal to unity, is now necessary:

$N_u = 0.57 \times 30 \times 400 \times 250 + 4700 \times 400/1.15$

$\quad = 3340$ kN

$N_u - N = 3340 - 1125 = 2215$ kN

$N_{bal} = 0.4 \times (30/1.5) \times 400 \times 250 = 800$ kN

$N_u - N_{bal} = 2540$

$k_2 = 2215/2540 = 0.87$

$M_s = 0.87 \times 49.15 = 42.76$ kN m

The variation in M_s is very small; hence the above value of $A_s = 4700$ mm² is acceptable. Provide six bars 25 mm diameter on each of the short edges of the column section at 40 mm embedment.

6.6.6 Column shear reinforcement

The design process for shear in columns (Figure 6.15) according to EC2 normally requires the following steps:

1. First obtain the design shear force acting on the section V_{Sd}.
2. Evaluate the minimum shear force V_{Rd1} that can be sustained by the section without any shear reinforcement. V_{Rd1} is given by

    $V_{Rd1} = [\tau \, k \, (1.2 + 0.4\rho) + 0.15\sigma]bh'$

    where $\tau = 0.035(f_{ck})^{2/3}$, $k = 1.6 - h' \nleq 1$ where h' is in metres, $\rho = 100A_{st}/bh' \ngtr 2$ and $\sigma = N/A_c$ $> 0.4f_{ck}/1.5$. A_{st} is the tension reinforcement
3. Evaluate the maximum shear force V_{Rd2} that the section can sustain without crushing the concrete. V_{Rd2} is given by

    $V_{Rd2} = 0.75 \, \nu \, [(f_{ck}/1.5) - \sigma_{eff}]bh'$

    where $\nu = (0.7 - f_{ck}/200) \nleq 0.5$ where f_{ck} is in N/mm² and $\sigma_{eff} = (N - A_{sc}f_{yck}/1.15)/A_c \nleq$ $0.4(f_{ck}/1.5)$. A_{sc} is the compression reinforcement.

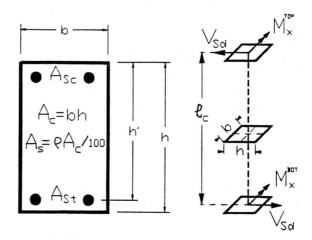

Figure 6.15 Analysis of rectangular concrete column with shear reinforcement.

4. If $V_{Sd} \leqslant V_{Rd1}$, nominal shear reinforcement only is required. If $V_{Rd2} \geqslant V_{Sd} \geqslant V_{Rd1}$, shear reinforcement required is given by

$$A_{sw} = (V_{Sd} - V_{Rd1})s/(0.9h' f_{ywk}/1.15)$$

where s is the pitch of the links and f_{ywk} is the steel links characteristic strength. If $V_{Sd} \geqslant V_{Rd2}$, the section is too small and needs to be enlarged. It should be noted that normally only nominal shear reinforcement is required for most commonly loaded columns.

Typical example

$V_{Sd} = 112 \, \text{kN}$ design shear force

$N = 1125 \, \text{kN}$ design axial load

$b = 250 \, \text{mm}$ $h = 400 \, \text{mm}$ $h' = 360 \, \text{mm}$

$f_{ck} = 30 \, \text{N/mm}^2$

$f_{yck} = f_{ywk} = 400 \, \text{N/mm}^2$

$\rho = 2.1\%$ and $A_{st} = 0$

Minimum shear strength

$\tau = 0.035(30)^{2/3} = 0.338 \, \text{N/mm}^2$

$k = 1.6 - h' = 1.6 - 0.36 = 1.24$

$\sigma = N/A_c = 1125\,000/(250 \times 400) = 11.25$

$V_{Rd1} = [0.338 \times 1.24 \times (1.2 + 0.4 \times 0.0)$

$\qquad\qquad + 0.15 \times 11.25] \times (250 \times 360)$

$\qquad = 197.1 \, \text{kN}$

Use nominal stirrups since $V_{Sd} \leqslant V_{Rd1}$.

6.7 WALLS AND PLATES LOADED IN THEIR OWN PLANE

6.7.1 Introduction

The function of structural walls is primarily to resist horizontal forces acting in their own plane, provide overall stability to the building and ensure its non-sway status, thus rendering the P–Δ effects negligible. Walls are acted upon like columns by axial loads, bending moments and shears; as a general rule a column is deemed to be considered as a wall when the ratio of the larger to the smaller cross-sectional dimension exceeds 4.

6.7.2 'Truss' models for walls, corbels and deep beams

EC2 guidance on the analysis of walls is useful but rather too general. The linear elastic analysis is one of the methods recommended, with the added hints that truss-like idealizations of walls are a possible way for the analysis; this is quite adequate for low-ductility designs. For high-ductility designs, there are often regions of high stress concentration or concrete cracking; the Code requires this to be taken into account and suggests that a reduction in the rigidity of the elements in the areas in question is an acceptable method to be employed in using the elastic analysis.

Walls used in buildings come in different shapes and forms and as such are difficult to analyse; idealizations normally have to be made, which are not necessarily universally agreed upon. Corbels and deep beams are structural forms frequently met in the design of buildings which are not amenable to exact or easy analyses. Openings of various descriptions can be present in walls and deep beams in the form of doors or service holes; stress concentrations are inevitable at the corners of openings and should be suitably reinforced using diagonal bars positioned at the corners of the openings.

EC2 recommends the use of 'truss'-like models in an elastic analysis as a realistic approximation for the analysis of walls, corbels and deep beams. To be efficient, any such analyses should be computer-oriented. Finite elements of all types and shapes are available for the analysis of such structures, and computer programs are easily accessible.

The authors have been using finite elements for the analysis of in-plane plate structures for some time, and a 'truss finite element' that can be incorporated with ease in any skeletal plane frame analysis computer program is described below with advice as to suitable idealizations for walls, corbels and deep beams with or without holes.

The finite element described in Figure 6.16 is based on the work of Hrennikoff (1941). The method was further developed for computer use by Yettran and Husain (1966).

EC2 recommends – in the absence of better data – that the permitted stress in the struts should be taken as $0.4f_{ck}/1.5$, and in the ties, where the entire force is taken by the steel reinforcement, the stress should be taken as $0.87f_{yk}$.

Poisson's ratio for concrete has been taken as 0.15. Members marked as 1 and 2 are taken as being fixed at both ends. For these

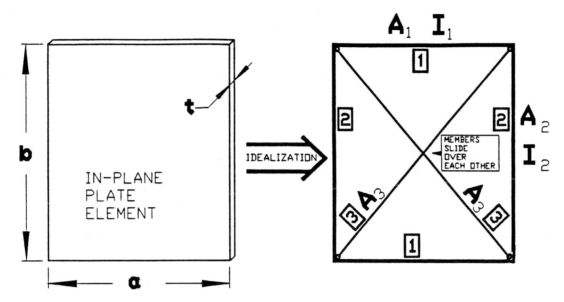

Figure 6.16 A 'truss finite element' for use in any computer package.

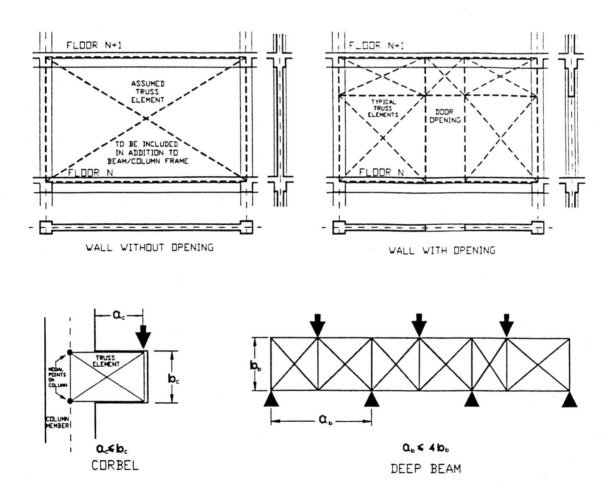

Figure 6.17 Typical truss idealizations.

$$A_1 = [1.023 - 0.153(a/b)^2](bt/2)$$

$$I_1 = 0.281(tb^3/12)$$

$$A_2 = [1.023 - 0.153(b/a)^2](at/2)$$

$$I_2 = 0.281(ta^3/12)$$

Members marked as 3 are taken as being pin jointed at both ends. For these

$$A_3 = [0.153d^2/(ab)](dt/2)$$

where $d = \sqrt{(a^2 + b^2)}$

The ratio of the sides a and b should lie within the limits: $2 \geqslant a/b \geqslant 0.5$.

Typical truss idealizations are shown in Figure 6.17.

6.7.3 First-stage wall design

The stresses in the wall shown in Figure 6.18, subject to an axial load N and an in-plane moment M, is given by

$$\sigma = N/(L_w b_w) \pm M/(b_w L^2{}_w/6)$$

The ultimate compressive stress σ_{max} at the compressive edge of the wall should be

$$\sigma_{max} < 0.57 f_{ck} + [A_{sc}/(L_w b_w)] (0.87 f_{yck})$$

where $A_{sc}/(L_w b_w)$ is the compression steel reinforcement ratio.

The tension steel reinforcement ratio $A_{st}/(L_w b_w)$, if required, is given by

$$A_{st}/(L_w b_w) > 0.5 (L_t/L_w)[\sigma_{max}/(0.87 f_{yk})]$$

It is advisable to place the derived tension steel area in the $0.5L_t$ region from the edge of the wall where the maximum tensile stress has developed.

6.8 REINFORCEMENT AND DIMENSIONAL LIMITS

6.8.1 Steel reinforcement limits (columns)

The minimum percentage longitudinal reinforcement should be equal to $17.25N/(bhf_{yk}) \not< 0.3$. The maximum reinforcement in the longitudinal direction, even at laps, should not exceed 8%, for cast *in situ* columns.

The minimum longitudinal bar diameter should not be less than 12 mm.

The minimum diameter of the links should not be less than 6 mm or one-quarter of the largest longitudinal bar diameter. The pitch of the links should be the smallest of the least column sectional dimension, 12 times the smallest longitudinal bar diameter and 300 mm. The pitch should be reduced to 0.6 times the one in the previous sentence, above and below beam column joints for a distance equal to the largest sectional dimension of the column, and at column laps if the longitudinal steel diameter exceeds 14 mm.

6.8.2 Dimensional limits (columns)

The minimum cross-sectional dimension for a vertically cast *in situ* column should not be less than 200 mm.

Tolerances in cross-sectional dimensions and cover should be borne in mind when calculating the reinforcement steel area; the possible error in the cover (EC2) that can occur when fixing the steel in cast *in situ* section lies between ±5 mm and ±10 mm. The errors in the cross-sectional dimensions can be taken as ±5 mm for linear dimensions up to 150 mm, linearly varying to ±15 mm at 400 mm and then to ±30 mm at 2500 mm.

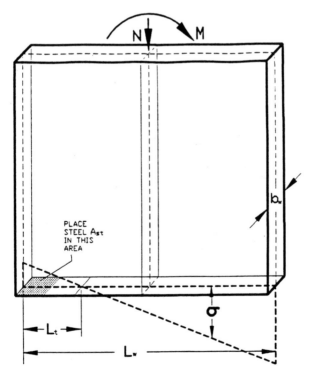

Figure 6.18 First-stage wall design.

6.8.3 Steel reinforcement limits (walls)

The minimum area of reinforcement should not be less than 0.4% of the cross-sectional area of the wall and should normally be applied shared equally on both faces of the wall.

The maximum longitutinal reinforcement should not exceed 4% of the cross-sectional area of the wall, and when it reaches 2% it should be adequately restrained transversely by closed stirrups.

The amount of the horizontal reinforcement should not be less than half the vertical steel, its diameter should not be less than one-quarter of the longitudinal bars, and the spacing should be limited to the smaller of 300 mm or twice the wall thickness b_w.

6.8.4 Dimensional limits (walls)

The length of the wall L_w should be greater than or equal to four times the width b_w. It is advisable for the minimum wall thickness not to fall below 150 mm.

6.9 EC8 (DRAFT) ADDITIONAL COLUMN DESIGN REQUIREMENTS

6.9.1 Introduction

The EC2 design approach described so far deals primarily with low-ductility designs. In earthquake regions, however, the designs need to be of medium or high ductility to allow the structure the capability of higher deformations with the formation of plastic hinges, prior to collapse, developing in the beams only, barring minor exemptions, and not in the columns. To achieve the above end, the column actions are suitably amplified prior to proceeding with the section design for bending and shear illustrated in the above sections.

6.9.2 Column moment amplification

Amplification of moments depends on the amplification factors α_{cd} or ζ_m and ζ_r given below (Figure 6.19):

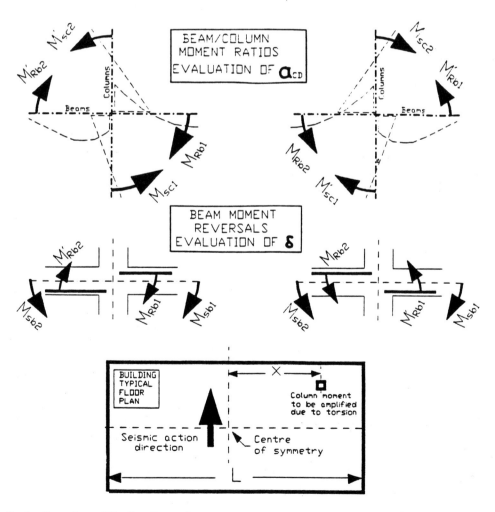

Figure 6.19 Evaluation of amplification factor ζ.

- The beam column moment ratios, from which the factor α_{cd} is evaluated.
- The moment reversal factor δ.
- The bending moment amplification is given by $\zeta_m = |1 + (\alpha_{cd} - 1) \delta|$.
- An additional amplification $\zeta_r = 1 + 0.6x/L$ to the bending moments is used to cater for the position of the column in the floor plan and its assumed contribution to resisting the torsional effects imposed on the structure due to the seismic actions.

The coefficient α_{cd} is given by the maximum of the following beam/column moments ratios:

$$\alpha_{cd} = \gamma_{Rd}(|M_{Rb1}| + |M'_{Rb2}|)/(|M_{Sc1}| + |M'_{Sc2}|)$$

$$M_{design} = \alpha_{cd} M_{sd} \quad \text{if } \alpha_{cd} \leq \zeta_m$$

or

$$\alpha_{cd} = \gamma_{Rd}(|M'_{Rb1}| + |M_{Rb2}|)/(|M'_{Sc1}| + |M_{Sc2}|)$$

where M_{sd} is the design acting moment, and where $\gamma_{Rd} = 1.35$ and $\gamma_{Rd} = 1.20$ for high- and medium-ductility designs respectively.

The moment reversals factor δ is the larger of the values obtained from the expressions below:

$$\delta = (|M_{Sb1} - M'_{Sb2}|)/(|M_{Rb1}| + |M'_{Rb2}|)$$

$$M_{design} = \zeta_m M_{sd} \quad \text{if } \zeta_m \leq \alpha_{cd} \text{ and } \zeta_m \leq q$$

or

$$\delta = (|M'_{Sb1} - M_{Sb2}|)/(|M'_{Rb1}| + |M_{Rb2}|)$$

$$M_{design} = q M_{sd} \quad \text{if } \zeta_m \leq \alpha_{cd} \text{ and } q \leq \zeta_m$$

6.9.3 Column shear amplification

The design shear forces should be evaluated using the actual longitudinal column reinforcements at the top and bottom of the column according to the expression below:

$$V_c = \gamma_n(M_{Rc,top} + M_{Rc,bot})/l_c \quad \text{where } \gamma_n = \gamma_{Rd}$$

6.9.4 Evaluation of moment amplification factors

(a) Top of the column

The column considered is shown in Figure 6.20. Assume that the column location amplification factor $\zeta_r = 1.10$. Then

capacity of beams as per reinforcement

$$= 70 + 10 = 80 \text{ or } 77$$

corresponding column capacity as per analysis

$$= 21 + 20 = 41 \text{ or } 66$$

$$\gamma_{Rd} = 1.2$$

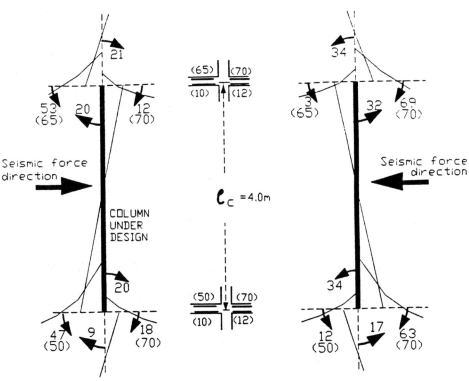

Figure 6.20 Analysis of moment amplification factors of column.

For medium ductility $\gamma_{Rd} = 1.2$ and $q = 3.0$. Then

maximum $\alpha_{cd} = 1.2 \times 80/41 = 2.34$

maximum beam moment reversal as per analysis

$= 69 - 12 = 57$

minimum capacity of beams

$= 65 + 12 = 77$

$\delta = 57/77 = 0.74$

$\zeta_m = 1 + (\alpha_{cd} - 1) \, \delta = 1.99 < q$

column maximum design moment

$= 1.1 \times 1.99 \times 32 = 70 \, \text{kN m}$

column minimum design moment

$= 1.1 \times 1.99 \times 20 = 43.8 \, \text{kN m}$

(b) Bottom of the column

In this case

capacity of beams as per reinforcement

$= 70 + 10 = 80 \text{ or } 62$

minimum column capacity as per analysis

$= 9 + 20 = 29 \text{ or } 51$

maximum $\alpha_{cd} = 1.2 \times 80/29 = 3.31$

maximum beam moment reversal as per analysis $= 63 - 18 = 45$

minimum capacity of beams

$= 50 + 12 = 62$

$\delta = 45/62 = 0.73$

$\zeta_m = 1 + (\alpha_{cd} - 1) \, \delta = 2.69 < q$

column maximum design moment

$= 1.1 \times 2.69 \times 34 = 100.6 \, \text{kN m}$

column minimum design moment

$= 1.1 \times 2.69 \times 20 = 59.2 \, \text{kN m}$

(c) Column design shear force

We have

$V_{cd,max} = 1.2 \times (70 + 100.6)/4.0 = 51.18 \, \text{kN}$

$V_{cd,min} = 1.2 \times (43.58 + 59.2)/4.0 = 30.9 \, \text{kN}$

where the column length in this case is 4 m.

6.9.5 Beam versus column plastic hinge formation

Beam hinge mechanisms are allowed in EC8 and almost all other Codes around the world, because they are more efficient in dissipating seismic energy compared to column hinge mechanisms. The basic reason for this choice is that beam hinges are more ductile than their column counterparts because they are required to carry lower axial loads; in addition it is the columns that are the components needed to sustain the high axial loads after the earthquake

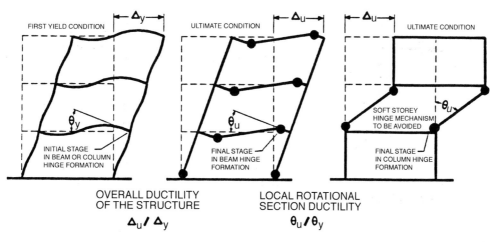

Figure 6.21 Plastic hinge formation.

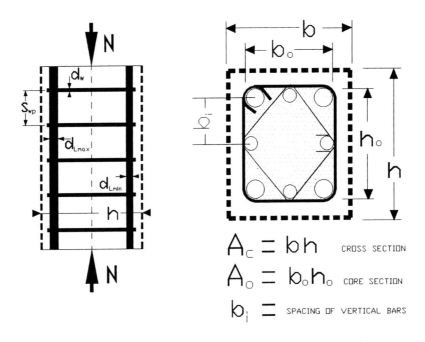

Figure 6.22   Confinement of concrete core using transverse links.

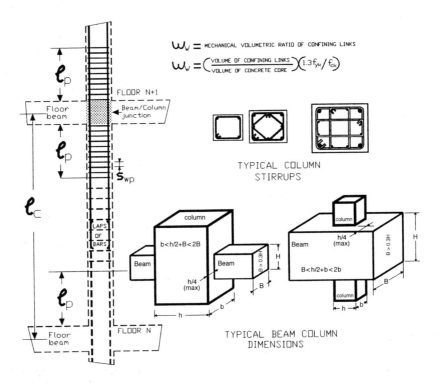

Figure 6.23   Details of 'plastic regions' of columns.

Table 6.5 Parameters in column 'plastic regions'

Ductility level, DL	Plastic region, l_p	Pitch, s_{wp}	Link's diameter, d_w	Volumetric links ratio, ω_w	Max., min. %ρ steel; and max. b_i	Column min. dim., $b < h$	Axial ratio, $\nu_{d,max}$
H	Largest of 1.5h $l_c/5$ 600	Smallest of $b_o/4$ $5d_{L,min}$ 100	Not less than $0.4d_{L,max}\sqrt{(f_{yk}/f_{ywk})}$ and $6 \leq d_w \leq 12$	$0.15 \leq \omega_w$	$1.0 \leq \rho \leq 4.0$ 150	300	0.55
M	Largest of 1.5h $l_c/6$ 450	Smallest of $b_o/3$ $7d_{L,min}$ 150	Not less than $0.35d_{L,max}\sqrt{(f_{yk}/f_{ywk})}$ and $6 \leq d_w \leq 12$	$0.07 \leq \omega_w$	$1.0 \leq \rho \leq 4.0$ 200	250	0.65
L	Largest of h $l_c/6$ 450	Smallest of $b_o/2$ $9d_{L,min}$ 200	$6 \leq d_w \leq 12$	$0.04 \leq \omega_w$	$1.0 \leq \rho \leq 4.0$ 250	200	0.75

comes to an end, thus preventing the structure from totally collapsing. It should also be noted that from the practical point of view it is more difficult to detail columns for high ductility than beams; column hinge mechanisms require the columns to develop higher ductilities, while beam hinge mechanisms occur at lower beam ductilities, as illustrated by Figure 6.21.

6.9.6 Transverse reinforcement function

During a severe earthquake, cross-sections of members in highly stressed areas (such as plastic hinge positions in beams and in column regions above and below the beam column joints) will be susceptible to spalling of the concrete cover due to reversal effects of the seismic actions. To allow for strength reduction due to the possible spalling of concrete, it is necessary to confine the concrete core of the cross-section using transverse links as shown in Figure 6.22. These links have a threefold function:

1. to cater for high shear forces;
2. to confine the concrete core, thus enhancing its longitudinal compressive strength;
3. to ensure that the exposed longitudinal steel reinforcement does not buckle.

6.9.7 Column plastic regions

The seismic design of columns requires that special care should be taken in reinforcing the column, above and below the floor beams framing into the column, in order to ensure that no hinges can develop in those areas of the column, but to equip the column with the ability to deform as would be required by a hinge formation in the beams. In order to cater for this essential column property, it is necessary to *confine* the concrete by providing sufficient lateral reinforcement in the form of links, in the regions above and below the joint, called 'plastic regions' of the column; the diameter and pitch of the links and other details are shown in Figure 6.23 and Table 6.5.

6.9.8 Evaluation of confining links reinforcement

The calculation of ω_w, as required by EC8, is rather cumbersome, as there are a variety of empirical cofficients that correspond to various ductility levels and links types. The computer program below (in QBasic) evaluates ω_w, A_{sw} and d_w for the maximum permitted pitch s_{wp}.

Computer program
See computer program 3.

6.10 EC8 (DRAFT) ADDITIONAL WALL DESIGN REQUIREMENTS

6.10.1 Introduction

EC8 deals with structures subject to earthquakes which should be capable of resisting such actions that may be exerted upon them due to severe ground shaking. Walls are essential elements that can, if properly reinforced, possess excellent energy dissipating properties to resist large seismic actions

```
CLS
t2 = 0.25: t1 = 0.15
INPUT "Concrete strength fck in N/sq.mm = "; fck
INPUT "Steel link strength fywk in N/sq.mm = "; fywk
INPUT "Main steel strength fyk in N/sq.mm = "; fyk
INPUT "Design axial load N in kN = "; N
INPUT "Minimum section dimension b in mm = "; b
INPUT "Maximum section dimensiom h in mm = "; h
INPUT "Cover to main steel c in mm = "; c
INPUT "Maximum main steel diameter dLmax in mm = "; dLmax
INPUT "Minimum main steel diameter dLmin in mm = "; dLmin
PRINT "Single link type enter 1"
PRINT "Double link type enter 2"
PRINT "Treble link type enter 3"
PRINT "High ductility enter 1"
PRINT "Medium ductility enter 2"
PRINT "Low ductility enter 3"
INPUT "Spacing of links in mm = "; swp

10      INPUT "Link type = "; lt: lt = INT(lt)
        IF lt > 3 OR lt < 1 THEN 10
20      INPUT "Ductility level = "; dl:
        IF INT(dl) > 3 OR INT(dl) < 1 THEN 20
        IF dl = 1 THEN
                IF bo/4 < 6 * dLmin THEN s = (b – 2 * c)/4 ELSE s = dLmin
                IF s > 100 THEN s = 100
                IF swp > s THEN swp = s
        END IF
        IF dl = 2 THEN
                IF bo/3 < 8 * dLmin THEN s = (b – 2 * c)/3 ELSE s = dLmin
                IF s > 150 THEN s = 150
                IF swp > s THEN swp = s
        END IF
        IF dl = 3 THEN
                IF bo/2 < 10 * dLmin THEN s = (b – 2 * c)/2 ELSE s = dLmin
                IF s > 200 THEN s = 200
                IF swp > s THEN swp = s
        END IF
IF INT(lt) = 1 AND INT(dl) = 1 THEN PRINT "Single link NOT suitable": GOTO 20
IF INT(lt) = 2 AND INT(dl) = 1 THEN lamda = 1.3
IF INT(lt) = 3 AND INT(dl) = 1 THEN lamda = 1.1
IF INT(lt) = 1 AND INT(dl) = 2 THEN lamda = 1.65
IF INT(lt) = 2 AND INT(dl) = 2 THEN lamda = 1.0
IF INT(lt) = 3 AND INT(dl) = 2 THEN lamda = 1.0
IF INT(lt) = 1 AND INT(dl) = 3 THEN lamda = 1.45: t1 = 0.1: t2 = 0.37
IF INT(lt) = 2 AND INT(dl) = 3 THEN lamda = 0.75: t1 = 0.15: t2 = 0.45
IF INT(lt) = 3 AND INT(dl) = 3 THEN lamda = 0.75: t1 = 0.15: t2 = 0.45
con = lamda * t1 * (b – 2 * c) * (h – 2 * c)/(b * h) – t2 * lamda
m = 1500 * lamda/(b * h * fck): ww = con + m * N
IF dl = 1 OR dl = 2 AND ww < 0.2 THEN ww = 0.2
IF dl = 3 AND ww < 0.1 THEN ww = 0.1
Vc = (b – 2 * c) * (h – 2 * c) * swp
Lsw1 = 4 * ((b + h)/2 – 2 * c): REM linktype 1
Lsw2 = Lsw1 + SQR((0.5 * b – c)^2 + (0.5 * h – c)^2): REM linktype 2
Lsw3 = 2 * Lsw1 + 2 * (b + h)/3: REM linktype 3
IF lt = 1 THEN Asw = (ww * Vc * fck)/(1.3 * Lsw1 * fywk)
IF lt = 2 THEN Asw = (ww * Vc * fck)/(1.3 * Lsw2 * fywk)
IF lt = 3 THEN Asw = (ww * Vc * fck)/(1.3 * Lsw3 * fywk)
dw = SQR(Asw/1.27)
CLS
```

Computer program 3

```
IF dw < 6 THEN dw = 6
IF dw > 12 THEN PRINT "***Pitch should reduce***"
IF dl = 1 THEN
        d = 0.4 * dLmax * SQR(fyk/fywk)
        IF dw < d AND d > 6 THEN dw = d
END IF
IF dl = 2 THEN
        d = 0.35 * dLmax * SQR(fyk/fywk)
        IF dw < d AND d > 6 THEN dw = d
END IF
PRINT "*****OUTPUT*****"
PRINT "Concrete strength in N/sq.mm = "; fck
PRINT "Steel strength in N/sq.mm = "; fywk
PRINT "Section width in mm = "; b
PRINT "Section depth in mm = "; h
PRINT "Ductility level = "; dl
PRINT "Link type = "; lt
PRINT "Pitch in mm = "; INT(swp)
PRINT "ww = "; INT(ww * 1000 + 0.5)/1000
PRINT "Steel area in sq.mm = "; INT(Asw)
PRINT "Link's diameter in mm = "; INT(dw + 0.5)
END
```

Computer program 3 *contd.*

effectively. Higher ductility levels of design are employed and as a consequence a plastic wall design analysis approach is recommended. Shapes of walls in common use are of rectangular T, L and I types. Thicknesses of walls are governed by the Code requirements to ensure efficient concrete placing and fire protection. Shear and stability requirements may require increase in thickness.

Structural walls that resist seismic forces should, as far as possible, be distributed at the periphery of the building, be made to carry as much of the vertical loads as possible, avoid concentration of lateral loads on one or two walls only, as this may have severe foundations design implications, and ensure that their stiffness and location is such as to cater for the inelastic deformations to be evenly distributed.

Cantilever walls can be considered as beam and/or column. The loads are assumed to be applied at each storey, the wall's stability is ensured by the floors, and a plastic hinge is expected to form at the base, which should be designed to have an adequate rotational capacity. Well detailed cantilever walls exhibit excellent energy dissipating properties and are ideal structural components to resist earthquake forces. It is essential to ensure that the mode of the wall failure is of ductile flexural hinge nature, and not of brittle shear failure mode; to achieve this it may be necessary to overestimate the design shear force. Another problem to look out for is the possible instability of the compressive edge of the wall, and thus one must provide the necessary thickening if required, or reduce the spacing of the longitudinal main bars.

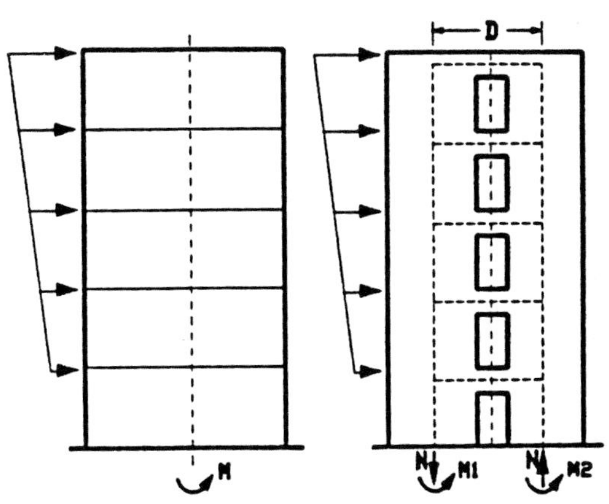

Figure 6.24 Comparison of flexural resisting mechanisms in structural walls.

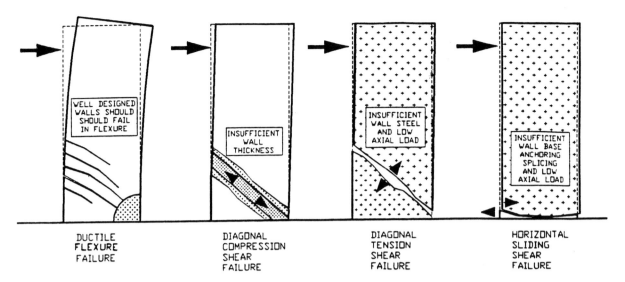

DUCTILE
FLEXURE
FAILURE

DIAGONAL
COMPRESSION
SHEAR
FAILURE

DIAGONAL
TENSION
SHEAR
FAILURE

HORIZONTAL
SLIDING
SHEAR
FAILURE

Figure 6.25 Modes of wall failure.

Structural walls with openings should be designed so that any plastic hinges formed should occur in the beams; hinge formation at the base of the wall only is permitted. Walls with openings are known as shear walls when the door openings appear in a regular pattern. Depending on the size of the openings, the engineer faces a dilemma: whether to consider it as one or two walls. Some guidance is given on this matter in Figure 6.24.

6.10.2 Modes of wall failure

Reinforced concrete wall failures (Figure 6.25) may be initiated by either flexure or shear. Shear failure could be due to diagonal tension or diagonal compression or could be of sliding type.

The following number of important facts should be borne in mind when designing reinforced concrete walls.

1. Cantilever walls are designed to form a plastic hinge at their base; the length of this hinge develops over a substantial length, sometimes being greater than the first storey height of the building, known as the critical length.
2. High concrete strains develop at the compressive edge, which could exceed the unconfined concrete strain of 0.0035; as a consequence the wall's ends should be confined with closed links in regions where the strain is greater than 0.0015.
3. To achieve large wall section ductilities, the majority of the main longitudinal steel reinforcement should be placed near the ends of the wall, with only nominal steel in the middle.

6.10.3 Wall section analysis

The method below was included in the draft stage of EC8 and provides a sound first-principles approach to the wall section analysis.

(a) Concrete stress–strain diagrams

The stress–strain diagram (Figure 6.26) for the wall section analysis consists of two parts. The first refers to the cross-sectional area of the wall subject to strains less than 0.0015 where the concrete is taken to be unconfined. The second part corresponds to strains above 0.0015 where the concrete is taken to be confined with stirrups at a specified pitch and of mechanical volumetric ratio $\alpha\omega_w \geq 0.1$ for medium and high ductilities.

(b) Wall's section moment curvature and strain diagrams

The basic assumptions for the evaluation of the required steel areas in walls is based on the specification that the design section's strain diagram (Figure 6.27) remains a straight line such that it can be adjusted by being moved parallel to itself to satisfy the two equations of equilibrium. This condition is achieved by the assumption that the sum of the extreme tensile steel strain e_{st} and the maximum compressive concrete strain e_{cu} remains constant, i.e. section's sum of strains:

$$e_{st} + e_{cu} = \text{constant}$$

Also curvature of the section:

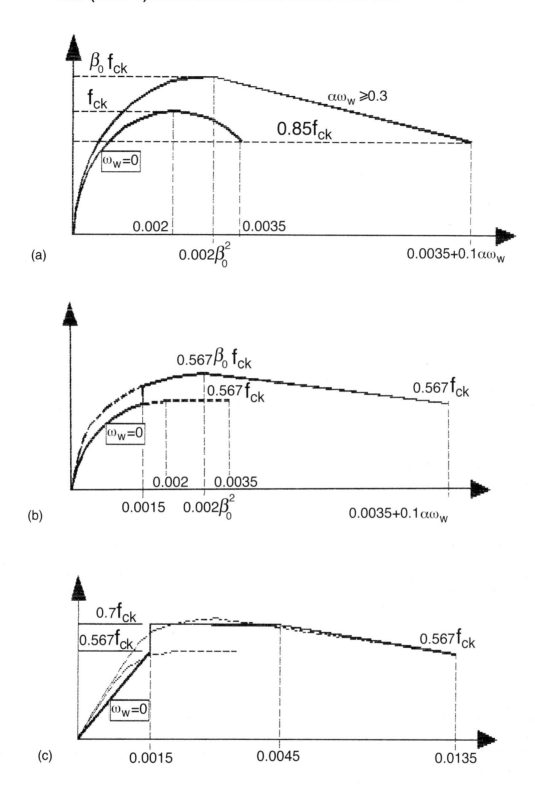

Figure 6.26 (a) Confined and unconfined concrete stress–strain diagram. (b) Design stress–strain diagram; factor $\beta_0 = 1.125 + 1.25\,\alpha\omega_w$. (c) Author's simplified stress–strain diagram.

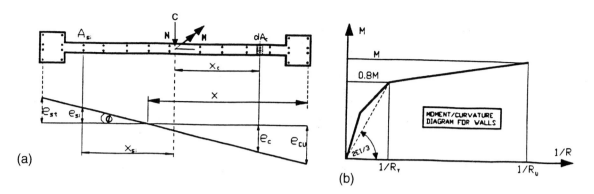

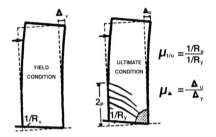

(a)

(b)

Figure 6.27   (a) Strain diagram and (b) moment–curvature diagram for wall section.

Figure 6.28   Local curvature ductility factor.

$$1/R_u = \varphi = (e_{st} + e_{cu})/L_w$$

and ultimate curvature $1/R_u$ in terms of curvature $1/R_y$ at yield:

$$1/R_u = \mu_{1/R} (1/R_y)$$

Whereas from the moment curvature diagram $1/R_y = 0.8M/(2EI/3)$, EC8 (Draft) conservatively specifies $1/R_y = 0.4M/EI$, and therefore $1/R_u = 0.4M\mu_{1/R}/EI$ where $\mu_{1/R}$ is the local curvature ductility factor.

(c)   Local curvature ductility factor $\mu_{1/R}$

The local ductility factor $\mu_{1/R}$ of the wall (Figure 6.28) depends on its overall deflection ductility factor μ_Δ, and $\lambda_p = L_p/L_w$, which is the ratio of the potential plastic hinge length to the overall length of the wall. In turn, μ_Δ is related to the behaviour factor of the structure q. Thus

$$q \leqslant \mu_\Delta \leqslant (q^2 + 1)/2$$

$$\mu_{1/R} = 1 + [(\mu_\Delta - 1)/(3\lambda_p^2 - 1.5\lambda_p)]$$

(d)   Wall's section equilibrium equations

Vertical equilibrium:

$$N = \int f_c \, dA_c + \sum (A_{si} \, f_{si})$$

where

$\int f_c \, dA_c$ and $\sum (A_{si} \, f_{si})$ are the resultant concrete and steel internal forces induced in the section under design.

Rotational equilibrium:

$$M = \int f_c x_c \, dA_c + \sum (A_{si} x_{si} f_{si})$$

Example 6.5:   wall section design example

The wall section for this example is shown in Figure 6.29.

Data and factors required

We have

minimum axial force = 3800 kN

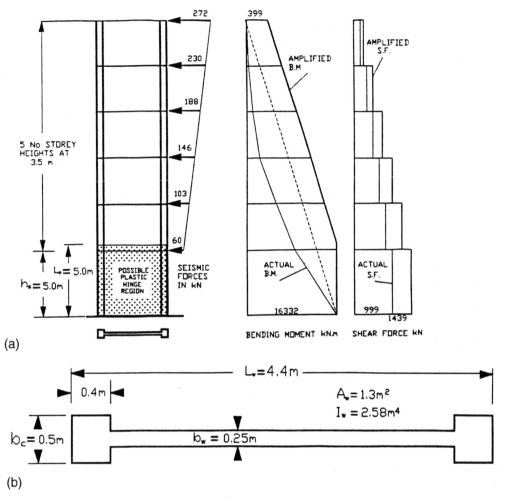

Figure 6.29 Wall section design example.

Shear ratio:

$$\alpha_s = M/(VL_w) = 16\,332/(1439 \times 4.4)$$

$$= 2.58, \quad \text{i.e. } \alpha_s > 1.3$$

Axial ratio:

$$ar = 1.5N_{min}/(A_w f_{ck}) = 1.5 \times 3.8 \times 10^6/(1.3 \times 10^6 \times 30) = 0.146$$

Wall slenderness:

$$h_w/L_w = 22.5/4.4 = 5.11$$

which is greater than 1.5, hence wall is slender. Steel reinforcement minimum ratio of wall:

$$\rho = A_s/A_w \geq 0.004$$

Local horizontal minimum steel reinforcement ratio:

$$\rho_h = A_h/(b_w S_v) > 0.0025$$

Local vertical minimum steel reinforcement ratio:

$$\rho_v = A_v/(b_w S_h) > 0.0025$$

Concrete shear stress:

$$\tau = 0.035 f_{ck}^{2/3} = 0.34 \text{ N/mm}^2$$

Concrete-to-concrete friction coefficient:

$$\mu_f = 0.33$$

Partial safety factor for medium-ductility level design:

$$\gamma_m = 1.3$$

Concrete Young's modulus:

$$E = 9.5 \times (40 + 8)^{1/3} = 34.5 \times 10^6 \text{ kN/m}^2$$

Diagonal compression failure mode check

$$V < 0.133f_{ck}A_w$$

$$V < 0.133 \times 40 \times 1.3 \times 10^3 = 6916 \text{ kN}$$

Diagonal tension failure mode check

$$V < (\rho_h f_{yhk}/1.15 + 2.5\tau) A_w$$

$$V < (0.0025 \times 400/1.15 + 2.5 \times 0.34) \times 1.3 \times 10^6$$

$$= 2235 \text{ kN}$$

Sliding shear failure mode check

$$V < [1.14\rho\sqrt{(f_{ck}f_{yk})}] A_w + \mu_f N_{min}$$

$$V < [1.14 \times 0.004\sqrt{(40 \times 400)}] 1.3 \times 10^6$$

$$+ 0.33 \times N_{min} = 2004 \text{ kN}$$

Hence design shear $V_d = 1439$ kN is below the least permissible value of $V_p = 2004$ kN. The ratio $V_p/(\gamma_m)V_d = 2004/(1.3 \times 1439) = 1.07$, which makes a flexural wall failure most probable, and, since the wall is slender ($h_w/L_w = 5.11$), flexural failure is assured.

It should be noted that the engineer must design walls whose modes of failure are at least of the most probable flexural failure type.

Values of μ_δ and μ_φ

We have

$$\mu_\delta < q \text{ and } \mu_\delta < (q^2 + 1)/2.$$

Therefore μ_δ is between 3.0 and 5.0.

Assume $\mu_\delta = 4.5$. Thus

$$\mu_{1/R} = 1 + (\mu_\delta - 1)/[3(L_p/h_w) - 1.5(L_p/h_w)^2]$$

$$= 6.91$$

Strain diagram line inclination

From Figure 6.30 we have

$$\varphi = 0.4 M\mu_{1/R}/EI$$

$$= 0.4 \times 16\,332 \times 6.91/(34.5 \times 2.58 \times 10^6)$$

$$= 0.000\,507 \text{ m}^{-1}$$

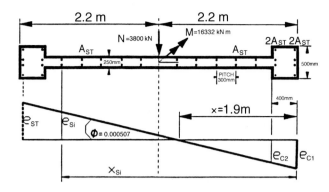

Figure 6.30 Inclination of strain diagram line.

$$e_{st} + e_{c1} = (L_w - c)\varphi$$

$$= (4.4 - 0.05) \times 0.000\,507$$

$$= 0.002\,21$$

We can use the simple computer program below to study the solution or trial and error to obtain the solution:

$$e_{c1} = 1.9 \times 0.000\,507 = 0.000\,96$$

(in the non-confined range)

$$f_{c1} = (0.000\,96/0.0015) \times 0.567 \times 40$$

$$= 14.52$$

$$f_{c2} = 14.52 \times 1.5/1.9 = 11.46 \text{ N/mm}^2$$

Evaluation of internal concrete force and moment

$$F_c = \int f_c \, dA_c = 4763 \text{ kN}$$

$$M_c = \int f_c x_c \, dA_c = 8103 \text{ kN m}$$

Evaluation of internal steel force and moment

We have

$$E_s = 200\,000 \text{ N/mm}^2$$

$$e_{sc} = 1.85 \times 0.000\,507$$

$$= 0.000\,94 \quad (\text{max. compression})$$

$$e_{st} = 2.45 \times 0.000\,507$$

$$= 0.001\,24 \quad (\text{max. tension})$$

$$e_{yd} = 400 \times 0.87/200\,000$$

$$= 0.001\,74 \quad (\text{in the elastic range})$$

```
REM DATA
FCK = 40: FYK = 400: FI = 0.000507
AW = 1300000: LW = 4.4

SCREEN 12
VIEW (1, 1)–(638, 450), 1, 2
WINDOW (0, 0)–(6, 20)

DIM xs(14)
REM STEEL AREAS POSITIONS
xs(1) = 50: xs(2) = 4050
xs(3) = 350: xs(4) = 4350
xs(5) = 650: xs(6) = 3750
xs(7) = 950: xs(8) = 3450
xs(9) = 1250: xs(10) = 3150
xs(11) = 1550: xs(12) = 2850
xs(13) = 1850: xs(14) = 2550

        FOR AST = 0 TO 3000 STEP 500

                FOR X = 400 TO 4000 STEP 100
                REM UNCONFINED CONCRETE
                ec = 0.0015: ec1 = FI * X/1000
                fc1 = 0.567 * FCK * ec1/ec: fc2 = fc1 * (X – 400)/X
                K1 = fc1 * X * 250/2
                K2 = (fc1 – fc2) * 400 * 250/2
                K3 = fc2 * 400 * 250

FC = K1 + K2 + K3
MC = K1 * (2200 – X/3) + ((2200 – 400/2)) * K2 + (2200 – 400/3) * K3

        REM STEEL STRESS BELOW FYK
                FS = 0: MS = 0: ESY = 0.87 * FYK/200000
                FOR i = 1 TO 14
                IF i <= 4 THEN
                AST1 = 2 * AST
                ELSE
                AST1 = 1 * AST
                END IF
                ESI = FI * (X – xs(i))/1000: FSI = 0.87 * FYK * (ESI/ESY)

FS = FS + AST1 * FSI
MS = MS + AST1 * FSI * (2000 – xs(i))

                NEXT

PRINT "FC, MC"; INT(FC)/1000, INT(MC)/1000000
PRINT "FS, MS"; INT(fs)/1000, INT(ms)/1000000

F = FC + FS: F = F/1000: F = 1000 * F/AW
M = MC + MS: M = M/1000000: M = 1000 * M/(AW * LW)

PRINT "AST, F, M"; INT(AST), INT(F), INT(M)

IF X = 400 THEN
PSET (M, F)
ELSE
LINE –(M, F)
END IF

        NEXT
NEXT

LINE (16332000/(AW * LW), 0)–(16332000/(AW * LW), 20), 4
LINE (0, 4000000/AW)–(20, 4000000/AW), 4
END
```

Computer program 4

Steel areas in the wall thickening are twice that in the wall web area. Thus

$$A_{si} = 1600 \text{ mm}^2$$

$$F_s = \sum A_{si}(E_s e_{si}) = -986 \text{ kN}$$

$$M_s = \sum A_{si}(E_s e_{si})(2.2 - x_i) = 8046 \text{ kN m}$$

and

total of internal forces = 3777 kN

total of internal moments = 16 149 kN m

total of external forces = 3800 kN

total of external moments = 16 332 kN m

To further study the above wall section design use can be made of the simple computer program 4.

REFERENCES AND BIBLIOGRAPHY

BCA (1993) *Concise Eurocode for the Design of Concrete Buildings*, BCA, UK.

CEC (1989) *Eurocode 8: Structures in Seismic Regions – Design, Part 1: General and Building* – Draft.

Hrennikoff A. (1941) Solutions of problems of elasticity by the framework method, *J. Appl. Mech. ASME*, **63**, December.

IAEE (1992) *Earthquake Resistant Regulations (1992), A World List*, International Association for Earthquake Engineering.

Newmark N.M. and Rosenblueth E. (1971) *Fundamentals of Earthquake Engineering*, Prentice-Hall, Englewood Cliffs, NJ.

Park R. (1986) Ductile design approach for reinforced concrete frames, *Earthq. Spect.*, **2**, No. 3, 565–614.

Park R. and Paulay T. (1975) *Reinforced Concrete Structures*, Wiley, New York.

Paulay T. (1970) An elasto-plastic analysis of shear walls, *ACI J.*, **67**, No. 11, 915–22.

Paulay T. (1972) Some aspects of shear wall design, *Bull. NZ Soc. Earthq. Eng.*, **5**, No. 3, 98–105.

Paulay T. (1986) The design of ductile reinforced concrete structural walls for earthquake resistance, *Earthq. Spect.*, **2**, No. 4, 783–832.

Paulay T. and Williams R.L. (1980) The analysis and design of and evaluation of design actions for reinforced concrete ductile shear wall structures, *Bull. NZ Nat. Soc. Earthq. Eng.*, **13**, No. 2, 108–43.

Steffens R.J. (1974) *Structural Vibration Damage*, BRE Report.

Yettran A.L. and Husain H.M. (1966) Plane framework methods for plates in extension, *J. Eng. Mech. Div. ASCE*, June.

7

EARTHQUAKE ENGINEERING

7.1 INTRODUCTION

Earthquakes have a great impact on people, their lives and property; the effects of devastation and loss of life during and after an earthquake are considerable. Despite our improved knowledge generally about earthquakes, their disastrous effects do not appear to diminish. Since 1980, to emphasize the point, 120 000 people have died in earthquakes worldwide, a figure that could and should be reduced if only the current technical understanding available were more effectively applied.

Earthquake engineering is a vast area of knowledge and is covered by a multitude of references, textbooks and specialist papers, some of which are listed in the 'references and bibliography' at the end of this chapter. The aim of this chapter is to concentrate on the fundamentals that an architect, engineer or other professional requires to know to effectively design reinforced concrete buildings in seismic areas.

An introductory explanation is given of earthquakes, where, why and how they occur, the means of their transmission and intensity measurement, and the development of seismic codes.

An attempt to explain how forces are transmitted to the building floor mass from the shaking of the ground is essential, for an enlightened design. A one-degree-of-freedom system and impulsive load analysis are given in order to introduce response spectra that are included in most national seismic codes, to provide an estimate of the acceleration induced in the floors of the building due to a given earthquake.

Multi-modal analysis of plane frames is given and the degree of participation of each higher mode is explained; the effect of damping is ignored in the modal analysis as its effect on the evaluation of the periods of vibration for concrete structures is normally small.

Real structures are mostly neither regular nor symmetrical as to either their mass or stiffness, and are constructed with frames, walls and/or shear walls to resist seismic forces. A multi-modal analysis for such structures is normally necessary and simplifying assumptions as to the number of degrees of freedom of the structure are required. It is normal in such cases to assume the floors of the buildings to be rigid in their own plane, thus reducing the number of degrees of freedom to three per floor: two linear displacements in the x and y directions and one rotation in the z direction. An earthquake analysis will give resultant forces on each floor of the building, which then need to be suitably shared amongst the constituent frame, wall and shear wall units, so that each can be analysed using standard software to obtain their design actions.

7.2 EARTHQUAKES

An earthquake is a sudden random motion, or trembling, in the Earth's upper crust caused by abrupt releases of accumulated strains in rocks below, volcanic activity, landslides and collapse of underground caves. The instrument that records the vibrations that take place during an earthquake is called

Figure 7.1 Types of fault movement evident after an earthquake.

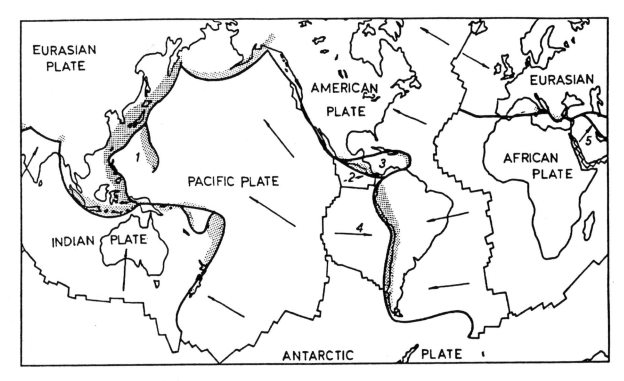

Figure 7.2 The Earth's crustal plates.

a **seismograph** and the record made is known as the seismogram. **Tectonics** is the branch of geology dealing with the structure of the upper part of the Earth's crust.

The Earth's crust is fractured in many places and rocks on either side of the fracture displace so that the strata no longer match. Such regions are known as **faults**, and the forces that cause movement to take place are compressive, tensile or shearing. They are shown in Figure 7.1 with associated movements evident after an earthquake.

The surface of the Earth is seen as being divided into a number of rigid plates, about 100 km thick, each of which 'floats' on some viscous underlayer. The crustal plates carry the continents and oceans, move relative to each other a few millimetres a year and interact at their boundaries. Most seismic activity is due to the interaction between these plates

and corresponds to the major active seismic belts along well-established fault lines. Figure 7.2 shows these plates diagrammatically.

7.2.1 Earthquake Transmission

Earthquake transmission or propagation is achieved through a number of waves initiated by the slipping and consequent energy release at the fault. Although the fault may have considerable length, it is commonly regarded as being generated from a single point source called the **focus**. The **epicentre** is the point on the Earth's surface *vertically above* the focus. There are two body and two surface waves associated with the transmission of earthquakes. The **body waves** are: (i) the P waves (primary or pressure waves) and (ii) the S waves (secondary or shear waves). The **surface waves** are: (i) the R waves (Rayleigh waves) and (ii) the L waves (Love waves).

The P waves are physically analogous to sound waves and travel with speed V_P of 5 to 13 km/s given by

$$V_P = (E/p)^{1/2}$$

where E and p are the Young's modulus and density, respectively, of the material through which the waves pass. They are longitudinal pressure waves causing a push–pull effect on the rock and/or soil particles as they pass through.

The S waves cause the particles of the material to move perpendicularly to the direction of propagation under a shearing action. The velocity of propagation V_S of S waves is 3 to 8 km/s, and thus slower than that of the P waves. The S wave speed is given by

$$V_S = (G/p)^{1/2}$$

where G is the shear modulus of the material through which the S waves pass.

The Love surface waves are propagated horizontally on the Earth's surface, with the ground movement being horizontally perpendicular to the direction of propagation. The velocity of propagation of these waves is slower than both the P and the S waves.

The Rayleigh surface waves are propagated on the Earth's surface in a horizontal direction on the surface with the ground movement being prominently in the vertical direction. The velocity of prop-agation of these waves is slower than the Love waves.

The Earth's crust, however, is not homogeneous, and seismic waves moving across interfaces of Earth materials of differing elastic properties cause changes in velocity and direction. A P wave, for example, when it encounters such an interface, gives rise to refracted, reflected P waves and/or refracted, reflected S waves (Figure 7.3).

7.2.2 Magnitude of an Earthquake

The magnitude of shallow-focus earthquakes is dependent on the energy released at the source of the disturbance. It is determined by measurements made on seismograms of an earthquake recorded at a seismic station 100 km from the epicentre and it is known as **Richter** M (Richter, 1935) on the magnitude scale:

$$\text{Richter } M = \log_{10}(A/A_0)$$

where A is the peak ground amplitude recorded by a Wood–Anderson seismograph and A_0 is the amplitude of one-thousandth of a millimetre. In practice, seismographs record at distances greater than 100 km and should be extrapolated to the standard distance from a number of recordings from different seismological stations. This value of Richter M is normally denoted as magnitude M_S. The Wood–Anderson seismograph becomes inaccurate for distances above 1000 km, and the Richter M measured within this distance is normally denoted

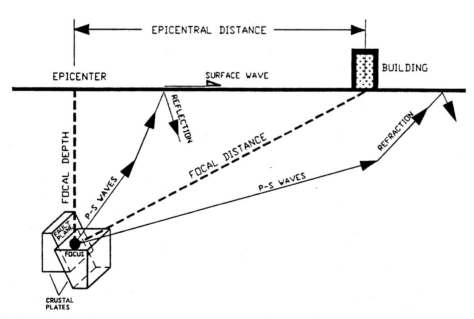

Figure 7.3 Refracted and reflected seismic waves in a non-homogeneous crust.

as magnitude M_L. Richter M based on body waves is denoted as M_b. It has to be appreciated that the measurement of M is a comparative and not a precise measure of the strength of an earthquake, owing to the anisotropy, viscosity, inhomogeneity and non-uniformity of the Earth's crust. The Richter M referred to in this chapter corresponds to M_S.

Complications arise in measuring the Richter M when two or more separate fault ruptures occur for which the ground shaking overlaps. Thus another magnitude scale, which claims to take into account multiple seismic events, known as **moment magnitude M_m**, was introduced by Hanks and Kanamori (1979):

$$M_m = (2/3)\log_{10}M_0 - 10.7$$

where $M_0 = GAD$

The seismic moment M_0 represents the energy release as a measure of the magnitude of the earthquake, where G is the shear modulus, A is the area of dislocation or fault surface and D is the average displacement of the slip on that surface. M_0, however, may be very difficult to evaluate.

7.2.3 Relationships between Richter *M* and energy release *E*

The empirical relationship below connects the energy release E during an earthquake and the Richter M:

$$\log_{10} E - \log_{10} E_0 = 1.5M$$

where $E_0 = 2.5 \times 10^{11}$ erg

The above formula has been verified by checking the seismic energy liberated by underground nuclear explosions. It is worth noting that the energy ratio between two earthquakes of Richter magnitudes M_1 and M_2 is given using the above formula by

$$E_1/E_2 = 10^{1.5(M_1-M_2)}$$

from which it can be seen that an increase in the Richter M of 1.0 corresponds to a 32-fold increase in the strength of the earthquake.

7.2.4 Intensity of an earthquake

The intensity of an earthquake is determined by the amount of damage or trembling caused by the disturbance. The intensity scale currently used in the Western world is the **modified Mercalli scale**, which

starts with intensity I, where the shock can only be detected by instruments, and extends to XII, where the destruction is total. To determine the intensities in a slight earthquake, the suspected area is normally canvassed by mail using specially prepared forms; for a strong earthquake, the area is visited by experts able to assess damage and assign intensities; from this data an isoseismal map may be drawn.

7.2.5 Relationships between Richter *M* and Mercalli *I*

Karnik (1969, 1971) proposed relationships between M and I for Europe and the eastern Mediterranean, which are as follows:

$$M = 0.50I + 1.8 \quad \text{Europe}$$

$$M = 0.44I + 2.43 \quad \text{eastern Mediterranean}$$

Gutenberg and Richter (1965) proposed a relationship for California, USA:

$$M = 0.67I + 1$$

Table 7.1 gives values using the above formulae for M and I, and approximate corresponding peak ground accelerations for comparison purposes.

7.2.6 Development of seismic codes

Seismic codes have evolved over the years and new provisions are only usually inserted after major earthquakes, when everyone has been shocked into action by the devastation and loss of life caused.

It was in 1909, after the Messina, Sicily, earthquake ($M = 7.5$), with many dead, that the Italian Commission specified that buildings should be designed to withstand a horizontal force $F = C_0W$,

Table 7.1 Values for M, I and approximate peak ground accelerations

Richter M			Modified Mercalli	Peak acc./
USA	Eur.	Med.	I	g
3.0	3.3	3.8	III	0.005
3.7	3.8	4.2	IV	0.01
4.4	4.3	4.6	V	0.02
5.0	4.8	5.1	VI	0.05
5.7	5.3	5.5	VII	0.1
6.4	5.8	6.0	VIII	0.2
7.0	6.3	6.4	IX	0.5
7.7	6.8	6.8	X	1.0

where $C_0 = 0.125$ and W is the total gravity load on the structure.

After the 1923 Kwanto, Japan, earthquake ($M = 8.3$), the Japanese adopted the design shear force $F = C_0 W$, where $C_0 = 0.1$

In 1933, and after many severe earthquakes, the Los Angeles City Code introduced the same formula for the horizontal seismic force $F = C_0 W$, but specifying a range of values for the seismic coefficient $0.08 \leqslant C_0 \leqslant 0.1$.

It was not until 1943 that it was realized that the *height* of the building had an influence on C_0; in 1957 a modification was introduced to take into account also the *flexibility* of the structure.

In 1959 the Structural Engineers Association, California, developed a seismic code for equivalent static forces that was in closer agreement with dynamic theory when *ductility* and the *natural frequency* of the structure were included. Thus the seismic force was now dependent on two parameters, i.e.

$$F = KCW$$

where

$$0.67 \leqslant K \leqslant 1.33$$

depending on the ductility of the structure, and

$$C = 0.05/T^{1/3}$$

where T is the natural frequency of the structure, which in turn depends on the height of the building as

$$T \simeq 0.1 \times (\text{number of storeys})$$

The distribution of the seismic force with the height of the building was assumed to be a line with maximum at the top storey and minimum at the bottom storey given by

$$F_i = [(W_i h_i)/(\Sigma W_i h_i)]F$$

This approach, with minor refinements and modifications, found its way into almost all national codes of practice around the world and is known as the **equivalent static analysis**.

7.3 EARTHQUAKE DYNAMIC ANALYSIS

7.3.1 Earthquake load

The seismic loading is caused by the ground acceleration $U_g(t)$, which is of appreciable random nature as shown in Figure 7.4.

7.3.2 Earthquake response

The response of the structure to an earthquake ground acceleration $U_g(t)$ is its resulting deflected form $U(t)$. The deflected form $U(t)$ of the structure shown in Figure 7.5 depends on a variety of factors such as its mass, stiffness, modes of vibration and energy absorption capability.

The aim of earthquake **dynamic analysis** is to find the means of obtaining $U(t)$ and hence $\ddot{U}(t)$ for the lumped masses of the structure, which are necessary in order to derive the storey forces that are induced as a result of the effects of the ground acceleration $\ddot{U}_g(t)$, i.e a relationship such as

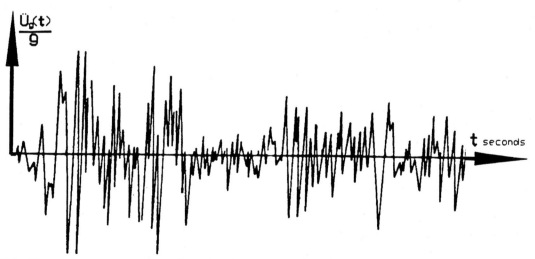

Figure 7.4 The random nature of ground acceleration in an earthquake.

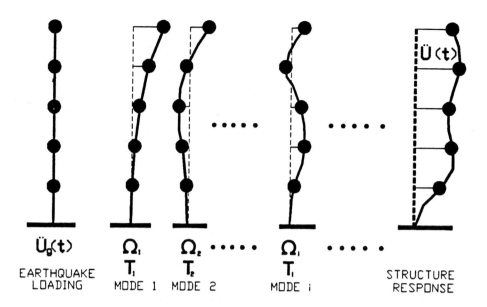

Figure 7.5 Response of a structure to an earthquake.

$$[\ddot{U}(t)/g]/[\ddot{U}_g(t)/g] = \Phi(M,k,\mu, T, q, S)$$

is sought, where Φ is a response transfer function, or simply the amplification factor to be applied to the ground acceleration. It depends on the mass M, stiffness k, natural period T, damping μ and ductility q of the structure, as well as the site soil conditions factor S. Most dynamic analyses for multi-degree-of-freedom structural systems currently in use utilize the single-degree-of-freedom system response transfer function.

In addition, although it is theoretically possible to obtain response transfer functions for single-degree-of-freedom systems, which take into account the non-linear behaviour due to ductility and site soil vibrations conditions, the required experimental data available are limited. As a consequence the current thinking is that the elastic response transfer functions that take into account damping μ and the structure's natural period T should be adjusted empirically to take into account the ductility and soil site foundation conditions. This is what is normally done in national seismic codes.

7.3.3 Single-degree-of-freedom system

Figure 7.6 shows a single-degree-of-freedom system before and during an earthquake. In the diagram, we have inertia force

$$F_I = (W/g) [\ddot{U}_g(t) + \ddot{U}(t)]$$

damping force

$$F_D = a_0 \dot{U}(t)$$

and elastic force

$$F_E = k(U(t)$$

From Newton's law

$$F_I + F_D + F_E = 0$$

$$(W/g)[\ddot{U}_g(t)] + \ddot{U}(t)] + a_0 U(t) + k\, U(t) = 0$$

$$\ddot{U} + (a_0 g/W)\, U + (kg/W)U = -\ddot{U}_g$$

Units of variables: W = weight of lumped mass, kN; W/g = mass (kg × 100), kN s²/m; a_0 = damping coefficient, kN s/m; k = structure stiffness, kN/m; $\ddot{U}_g$ = ground acceleration, m/s²; $\ddot{U}$ = acceleration of mass, m/s²; $\dot{U}$ = velocity of mass, m/s; U = displacement of mass, m.

The above dynamic equation of motion can be written as

$$\ddot{U} + 2\mu\Omega\, \dot{U} + \Omega^2\, U = -\ddot{U}_g$$

where

$$\mu = a_0/[(2W/g)(kg/W)^{1/2}]$$

Here μ is the damping ratio; for concrete $\mu = 0.01$ to 0.2. The normal value assumed in seismic codes is $\mu = 0.05$. Damping is known as 'critical' when $\mu = 1$. In addition,

$$\Omega = (kg/W)^{1/2} \quad \text{and} \quad T = 2\,\pi/\Omega$$

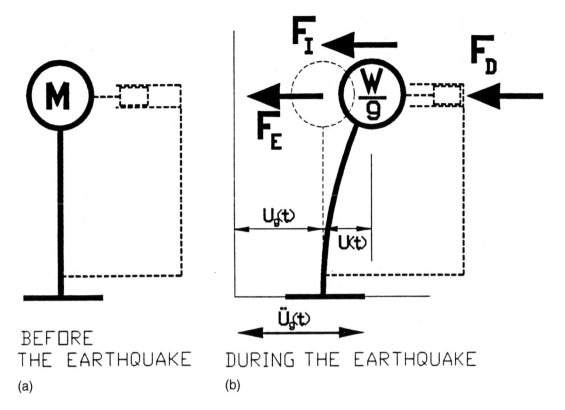

Figure 7.6 Single-degree-of-freedom system (a) before and (b) during earthquake.

where Ω is the undamped natural cyclic frequency in cycles per second for the structure and T its natural frequency in seconds.

When the vibration is free, the above differential equation can be written as

$$\ddot{U} + 2\mu\Omega\dot{U} + \Omega^2 U = 0$$

and its solution is

$$U(t) = e^{-\mu\Omega t} [A \sin(\Omega_D t) + B\cos(\Omega_D t)]$$

where $\Omega_D = \Omega(1-\mu^2)^{1/2} \simeq \Omega$ since $\mu < 0.2$.

The above solution can be simplified to

$$U(t) = A\, e^{-\mu\Omega t} \sin(\Omega_D t),$$

and when it is assumed that $t = 0$ and $U(0) = 0$, the solution can be represented by Figure 7.7.

The logarithmic decrement $\delta = \log_e [U(t+T)/U(t)]$ $= \mu\Omega T$ and therefore $\delta = 2\pi\mu$, which gives a means of evaluating the damping ratio from the logarithmic decrement.

7.3.4 Harmonic ground motion

Although the ground motion during an earthquake is of random nature, some useful pointers can be derived by considering the simple harmonic case,

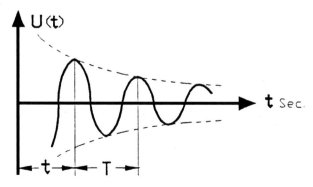

Figure 7.7 Damped solution $U(t)$ for condition $U(0) = 0$ at $t = 0$.

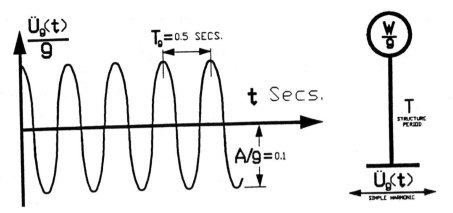

Figure 7.8 Typical simple harmonic ground motion.

where $\ddot{U}_g = \ddot{U}_{pg} \cos(2\pi pt)$ where $\ddot{U}_{pg}$ is the peak ground acceleration. Assume, for example, $\ddot{U}_{pg} = 0.1$ and $p = 2$ Hz. Then the ground acceleration versus time graph for the structure is given in Figure 7.8.

The single-degree-of-freedom response is given from the particular solution of the differential equation by the well-known expression

$$U(t) = \{-\ddot{U}_{pg}/[(\Omega^2 - (2\pi p)^2)^2 +$$

$$4((\mu\Omega)(2\pi p))^2]^{1/2}\} \cos(2\pi pt - \varphi)$$

where

$$\varphi = \tan^{-1}\{(2\mu\Omega)(2\pi p)/[\Omega^2 - (2\pi p)^2]\}$$

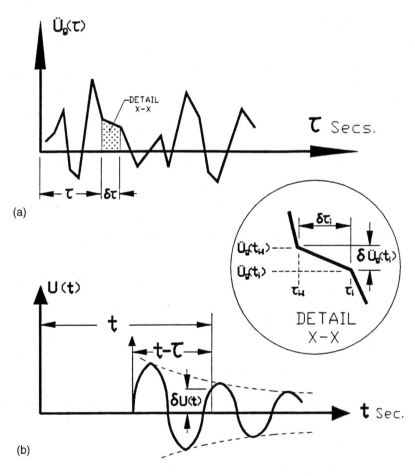

Figure 7.9 (a) Typical random ground motion and (b) response of single-degree-of-freedom system.

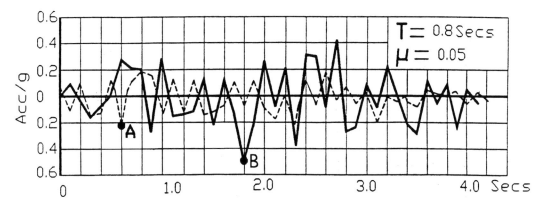

Figure 7.10 Computer output for single-degree-of-freedom system. The peak ground acceleration $A_{pg}/g = \alpha$ occurs at A, and the peak response acceleration $A_{pr}/g = 0.52$ occurs at B.

from which it can be seen, by differentiating $U(t)$ twice, and using the example in Figure 7.8, that the maximum mass acceleration occurs at $T^{-1} = p$, which is approximately the resonant condition, and that $\ddot{U}_{pr} \simeq (0.5/\mu)\ddot{U}_{pg}$; thus the amplification factor $\alpha = \ddot{U}_{pr}/\ddot{U}_{pg}$ is inversely proportional to the damping.

7.4 EARTHQUAKE RANDOM GROUND MOTION

The ground acceleration $\ddot{U}_g(\tau)$ due to an earthquake is a random function of time τ, and the corresponding response of a single-degree-of-freedom system deformation $U(t)$ is shown in Figure 7.9.

The elemental displacement $\delta U(t)$ due to any individual time step $\delta\tau$ of the ground acceleration $\ddot{U}_g(\tau)$ for the single-degree-of-freedom system is

$$\delta U(t) = [\ddot{U}_g(\tau)/\Omega_D] \, e^{-\mu\Omega(t-\tau)} \sin[\Omega_D(t-\tau)] \, \delta\tau$$

$$U(t) = (1/\Omega_D) \int \ddot{U}_g(\tau) \, e^{-\mu\Omega(t-\tau)} \sin[\Omega_D(t-\tau)] \, d\tau$$

This is known as the Duhamel integral and can be rearranged as follows:

$$U(t) =$$

$$+(e^{-\mu\Omega t}/\Omega_D) \sin[\Omega_D(t)] \int \ddot{U}_g(\tau) \, e^{\mu\Omega\tau} \cos[\Omega_D(\tau)] \, d\tau$$

$$- (e^{-\mu\Omega t}/\Omega_D) \cos[\Omega_D(t)] \int \ddot{U}_g(\tau) \, e^{\mu\Omega\tau} \sin[\Omega_D(\tau)] \, d\tau$$

$\ddot{U}_g(\tau)$ is expressed as a straight line between t_{i-1} and t_i and hence integration is carried out in a piecewise manner. A computer program can produce values of $U(t)$, $\dot{U}(t)$ and $\ddot{U}(t)$ for a given ground acceleration trace $\ddot{U}_g(\tau)$, a specified μ and T for the single-degree-of-freedom structure, which can be represented as shown in Figure 7.10.

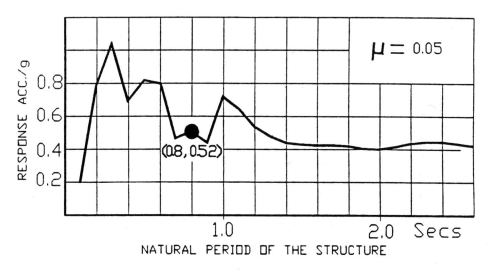

Figure 7.11 Typical acceleration response spectrum. The coordinates of the point marked are obtained from Figure 7.10.

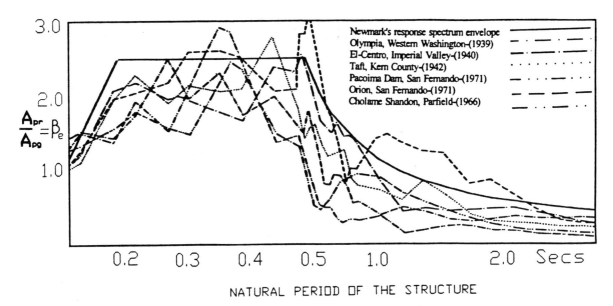

Figure 7.12 Producing a design acceleration response spectrum. After Newmark and Rosenblueth (1971).

7.4.1 Acceleration response spectrum

From the above analysis, using the Duhamel integral, it can be seen that the response transfer function $\Phi(T, \mu)$ can be obtained, connecting the ground peak acceleration A_{pg} to the peak mass acceleration A_{pr} for a given period T and damping μ for the single-degree-of-freedom system. The ratio $\beta_e(T, \mu) = A_{pr}/A_{pg}$ is the **response amplification factor**. The graph of A_{pr} versus T for a given μ can be plotted as in Figure 7.11. Such a plot is known as an **acceleration response spectrum** corresponding to the given seismic acceleration trace. Similar spectra for velocity and displacement can also be derived.

7.4.2 Design acceleration response spectrum

The above process can be repeated on a large number of known earthquake ground acceleration record traces and each individual response spectrum plotted on the same axes (Figure 7.12). A suitable envelope can then be chosen to replace the above plots. Such an envelope is known as a **design response spectrum**, and can be used in seismic design codes as a means of obtaining the mass response acceleration corresponding to a given natural period T of the structure. It should be noted, however, that single-degree-of-freedom response spectra are used, by seismic codes, to deal with multi-degree-of-freedom systems. Although this is not strictly correct, it is a necessary compromise owing to the complexity of producing multi-degree-of-freedom system response spectra.

7.4.3 EC8 design response spectrum

The EC8 design acceleration response spectrum $S(T) = \beta_e(T)(\alpha S/q)$ is shown in Figure 7.13.

7.4.4 Simplified dynamic analysis

In this approach no dynamic analysis is required; it is, however, necessary to derive the natural period T of the structure. This, in the absence of a more accurate value, is approximately given by $2\sqrt{\delta}$ seconds, where δ (m) is the lateral deflection at the top of the building due to gravity loads applied horizontally. Another easy-to-remember expression for the period is $T \simeq 0.1 \times$ (number of storeys in the building).

As mentioned before, for design purposes and in order to avoid the need for non-linear analysis, the concept of the behaviour and soil factors q and S was introduced in EC8. Thus the base shear (total horizontal force on the building) is given by:

$$F = \beta_e(T)(\alpha S/q)\ W = S(T)\ W$$

where α is the ground peak acceleration divided by the acceleration due to gravity g and W is the total gravity load acting on the building during the earthquake.

It should be noted that the Code's elastic response spectrum above caters for an adjustment in its shape via different values for T_C, which affect the starting point of its curved part as shown in Figure 7.13. The value of F is additionally amplified by an importance factor I, which for ordinary buildings is equal to unity.

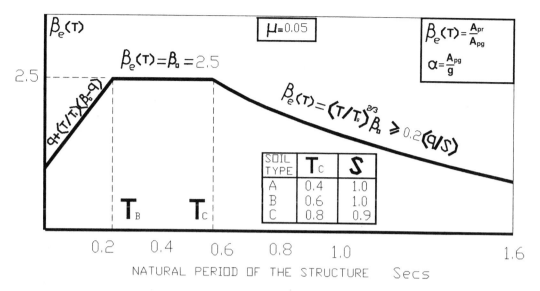

Figure 7.13 EC8 design response spectrum.

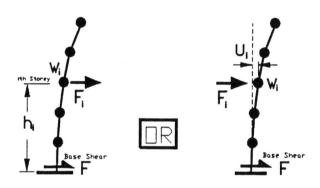

Figure 7.14 Distribution of base shear on each floor of a building.

Having evaluated the base shear, it only remains to distribute it suitably on each floor (Figure 7.14). This is done by using the distribution factors below:

$$F_i = [W_i \, h_i / \Sigma(W h)] \, F$$

or

$$F_i = [W_i \, U_i / \Sigma(W \, U)] \, F$$

where W_i is the ith storey's gravity load, h_i and U_i are the ith storey height from the foundation and the displacement of the ith floor respectively, $\Sigma(W h)$ is the sum of the storey weights times their heights from the foundation, and $\Sigma(W U)$ is the sum of the storey weights times the displacement of their corresponding floor.

7.5 MODAL ANALYSIS

The periods and modes of vibration of multi-degree-of-freedom systems can be obtained by considering the free undamped vibration, since it was established that, for single-degree-of-freedom concrete systems, the range of damping coefficients met makes $\Omega_D \simeq \Omega$.

The displacements and actions sequence chosen in Figure 7.15 is intentional to illustrate primary and secondary displacements.

7.5.1 Vibration equations

Let the mass of the structure be lumped at the nodes, and let the force and displacement vectors be U and p, where

$$U = \{U_1, \, U_2, \, U_3, \, \ldots, \, U_N\}^T$$

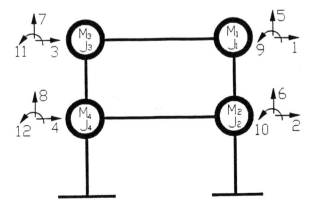

Figure 7.15 Sequence for a modal analysis.

and

$$p = \{p_1, p_2, p_3, \ldots, p_N\}$$

and where $\{U_1, U_2\}$ and $\{U_3, \ldots, U_N\}$ are the primary and secondary dynamic displacements respectively. By Newton's law

$$p = - M \ddot{U}$$

Using the stiffness method of analysis

$$p = KU$$

Assuming simple harmonic vibration

$$\ddot{U} = - \Omega^2 I U$$

where M, K and I are the mass, stiffness and unit matrices respectively.

Combining the three equations, the dynamic equation of motion is obtained as

$$M \ddot{U} + KU = 0$$

$$\ddot{U} + M^{-1} KU = 0$$

$$[M^{-1}K - (\Omega^2 I)]U = 0$$

$$[M^{-1}K - (2\pi/T)^2]U = 0$$

This is an eigenvalue problem; thus by solving $|M^{-1}K - (2\pi/T)^2| = 0$, the eigenvalues are obtained from the resulting characteristic polynomial. Note that the mode with the largest period of vibration T is the fundamental mode, and T is the natural period of the structure. The modes of vibration are the eigenvectors and are obtained by direct substitution into the equation

$$M^{-1}KU = (2\pi/T)^2 I \, U$$

of the eigenvalues $T_1, T_2, \ldots T_j, \ldots, T_N$ in turn. Thus

$$M^{-1}KU_1 = U_1(2\pi/T_1)^2$$

$$M^{-1}KU_2 = U_2(2\pi/T_2)^2$$

$$\vdots$$

$$M^{-1}KU_j = U_j(2\pi/T_j)^2$$

$$\vdots$$

$$M^{-1}KU_N = U_N (2\pi/T_N)^2$$

which in matrix notation can be written as

$$M^{-1}K\Phi = \Phi S$$

where Φ is the matrix that has the eigenvectors as its columns and is known as the **modal matrix**, and S is the diagonal matrix containing the periods $(2\pi/T_j)^2$ and is known as the **spectral matrix**.

It should be noted that the eigenvectors are orthogonal and therefore.

$$\Phi^T (M^{-1}K) \, \Phi = S$$

$$\Phi^T_N K \, \Phi = 0$$

$$\Phi^T_N M \, \Phi = 0$$

Let $U = \Phi \, Y$ where Y is known as the normalized displacement. Substituting into the equation of motion $M \ddot{U} + KU = 0$ for U and pre-multiplying by Φ^T, the following expression is obtained:

$$(\Phi^T M \, \Phi) \, \ddot{Y} + (\Phi^T K \, \Phi) \, Y = 0$$

which demonstrates that the modes of vibration are uncoupled, i.e. the deflected form of each mode can be treated separately as a single-degree-of-freedom system. The solution takes the form of N differential equations of the type

$$(\Phi^T_1 M \, \Phi_1) \, \ddot{Y} + (\Phi^T_1 K \, \Phi_1) \, Y = 0$$

$$(\Phi^T_2 M \, \Phi_2) \, \ddot{Y} + (\Phi^T_2 K \, \Phi_2) \, Y = 0$$

$$\vdots$$

$$(\Phi^T_j M \, \Phi_j) \, \ddot{Y} + (\Phi^T_j K \, \Phi_j) \, Y = 0$$

$$\vdots$$

$$(\Phi^T_N M \, \Phi_N) \, \ddot{Y} + (\Phi^T_N K \, \Phi_N) \, Y = 0$$

7.5.2 Earthquake response analysis

The differential equation of motion with damping and ground acceleration is given by

$$M\ddot{U} + C\dot{U} + KU = - M\ddot{U}_g$$

where M, C and K are the mass, damping and stiffness matrices, U is the matrix of primary displacements and $\ddot{U}_g$ the ground acceleration. It should be noted that in practice it is not possible to evaluate the damping matrix C and it is conventionally taken as: $C = 2\mu \, \Omega \, M$.

If the above equation is expressed in normalized coordinates, it can be stated as

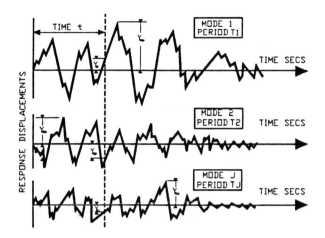

Figure 7.16 Response time histories of first, second, . . ., jth modes.

$$(\Phi^T M \Phi)\ddot{Y} + 2\mu\,\Omega\,(\Phi^T M \Phi)\,\dot{Y} + (\Phi^T K \Phi)\,Y$$

$$= -(\Phi^T_N M \hat{I})\ddot{U}_g$$

where $\hat{I}$ is a column vector of ones. Since the modes of vibration are orthogonal, the jth mode differential equation is

$$\ddot{Y}_j + 2\mu\,\Omega_j\,\dot{Y}_j + \Omega^2 Y_j = -[(\Phi^T_N M \hat{I})/(\Phi^T M \Phi)]\,\ddot{U}_g$$

where

$$[(\Phi^T_j M \hat{I})/(\Phi^T_j M \Phi_j)] = L_j$$

is known as the **participation factor**, and its denominator as the generalized mass of the $j$the mode of the vibration. The solution of the differential equation for the jth mode would yield the time history

response for the structure of period T_j and damping μ, using the Duhamel integral in a similar way to the single-degree-of-freedom system vibration:

$$Y_j(t) = -[(\Phi^T_j M \hat{I})/(\Phi^T_j M \Phi_j)]\,(1/\Omega_{Dj})$$

$$\int \ddot{U}_g(\tau)\,e^{-\mu\Omega_j(t-\tau)}\,\sin[\Omega_{Dj}\,(t-\tau)]\,d_\tau$$

where $\Omega_j \simeq \Omega_{Dj}$ for a concrete structure. The displacement response is $Y_j(t) = -L_j\,Y_{0j}$, where Y_{01}, Y_{02}, ..., Y_{0j} are the response time histories of the first, second, ... and jth modes shown in Figure 7.16. Similar time response histories can be obtained for the velocity and acceleration of the jth mode given by:

$$\dot{Y}_j(t) = -L_j\dot{Y}_{0j} \text{ and } \ddot{Y}_j(t) = -L_j\ddot{Y}_{0j}.$$

7.5.3 Participation factors evaluation

For the structure in Figure 7.17 we find

$$M = \begin{bmatrix} W_1/g & & & & \\ & W_2/g & & & \\ & & \cdot & & \\ & & & \cdot & \\ & & & & W_i/g \\ & & & & & \cdot \end{bmatrix}$$

$$\Phi = [\quad \Phi_1 \quad \Phi_1 \quad \ldots \quad \Phi_j \quad \ldots \quad]$$

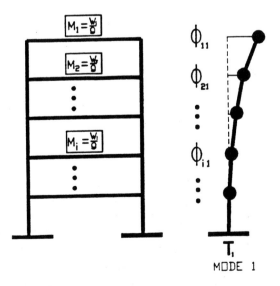
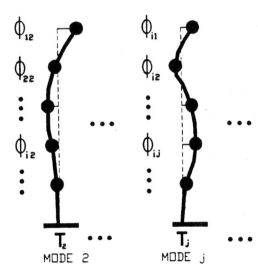

Figure 7.17 Structure used to evaluate participation factors.

$$\Phi = \begin{bmatrix} \Phi_{11} & \Phi_{12} & \ldots & \Phi_{1j} & \ldots \\ \Phi_{21} & \Phi_{22} & \ldots & \Phi_{2j} & \ldots \\ \cdot & \cdot & & \cdot & \\ \cdot & \cdot & & \cdot & \\ \Phi_{i1} & \Phi_{i2} & \ldots & \Phi_{ij} & \ldots \\ \cdot & \cdot & & \cdot & \\ \cdot & \cdot & & \cdot & \end{bmatrix}$$

$$\Phi^T_j\, M\, \hat{I} = [\Phi_{1j} \quad \Phi_{2j} \quad \ldots \quad \Phi_{ij} \quad \ldots]$$

$$\begin{bmatrix} W_1/g & & & & \\ & W_2/g & & & \\ & & \cdot & & \\ & & & \cdot & \\ & & & & W_i/g \\ & & & & & \cdot \\ & & & & & & \cdot \end{bmatrix}$$

$$\begin{bmatrix} 1 \\ 1 \\ \cdot \\ \cdot \\ \cdot \\ 1 \\ \cdot \\ \cdot \end{bmatrix} = \sum[\Phi_{ij}(W_i/g)]$$

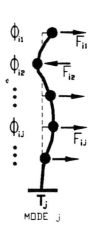

Figure 7.18 Structure used to evaluate inertial force vector.

The participation factor of the jth mode is $L_j = \Sigma[\Phi_{ij}(W_i/g)]/\Sigma[\Phi^2_{ij}(W_i/g)]$. Note that the generalized mass term $\Sigma[\Phi^2_{ij}(W_i/g)]$ can be set to unity by adjusting the values of jth mode vector Φ_{ij}. Thus the participation factor then takes the simplified form $\Sigma[\Phi_{ij}(W_i/g)]$, where $\Phi_{ij} = \Phi_{ij}/\sqrt{\Sigma[\Phi^2_{ij}(W_i/g)]}$ is the modified normalized jth mode vector.

7.5.4 Intertial force vector

The inertial force vector F_j, i.e. the seismic forces induced and acting on each of the lumped masses of the structure, corresponding to the jth mode at $t = t_0$ seconds, (Figure 7.18, can be obtained from the displacement time history by

$$\Phi^T_j\, M\, \Phi_j = [\Phi_{1j} \quad \Phi_{2j} \quad \ldots \quad \Phi_{ij} \quad \ldots]$$

$$\begin{bmatrix} W_1/g & & & & \\ & W_2/g & & & \\ & & \cdot & & \\ & & & \cdot & \\ & & & & W_i/g \\ & & & & & \cdot \\ & & & & & & \cdot \end{bmatrix}$$

$$\begin{bmatrix} \Phi_{1j} \\ \Phi_{2j} \\ \cdot \\ \cdot \\ \Phi_{ij} \\ \cdot \\ \cdot \\ \cdot \end{bmatrix} = \sum[\Phi_{ij}^2(W_i/g)]$$

$$F_j = (\Omega^2_j L_j Y_{0j}) M \Phi_j$$

$$= [(2\pi/T_j)^2 L_j Y_{0j}] \begin{bmatrix} M_1 \Phi_{1j} \\ M_2 \Phi_{2j} \\ \cdot \\ \cdot \\ M_i \Phi_{ij} \\ \cdot \end{bmatrix}$$

$$= \begin{bmatrix} F_{1j} \\ F_{2j} \\ \cdot \\ \cdot \\ F_{ij} \\ \cdot \end{bmatrix}$$

The resultant inertial forces on the lumped masses at $t = t_0$ seconds are obtained by summing up the intertial forces of all *effective modes*, which are normally not more than four for multi-storey buildings and they correspond to the larger periods of vibration.

The above method of obtaining the resultant interial force, although interesting, is not very useful, since the engineer needs to look into an infinite number of cases to find the worst case to use in the design.

It is therefore, as a compromise, assumed that the peak time history displacements, Y_{01p}, Y_{02p}, ... Y_{0jp}, ... are used. The difficulty, however, is that these peak values do not correspond to the same period of time $t = t_0$.

A general approximation in use, which appears in most seismic codes, is based on the square root of the sum of the squares of the inertial forces (Figure 7.19), obtained using the peak displacement values for each mode:

$$F_T = \begin{bmatrix} F_1 \\ F_2 \\ . \\ . \\ . \\ F_i \\ . \\ . \\ . \end{bmatrix} = \begin{bmatrix} \sqrt{(\Sigma\ F_{1j}^2)} \\ \sqrt{(\Sigma\ F_{2j}^2)} \\ \vdots \\ \sqrt{(\Sigma\ F_{ij}^2)} \\ \vdots \end{bmatrix}$$

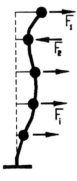

Figure 7.19 Simpler approximation to evaluate inertial force vector.

Similar expressions can be obtained for displacements, velocities and accelerations.

7.5.5 Inertial forces using the response spectrum

The approach for evaluating the inertial forces on the structure using the response spectrum is as follows.

1. Modal analysis should be carried out as above and the periods T and modes of vibration Φ established.
2. Evaluate the participation factors L for each mode of vibration.
3. Use the response spectrum to obtain the amplified response mass acceleration $\beta\ (T)$, as explained under sections 7.4.1 and 7.4.2, corresponding to the period T of each mode of vibration.
4. The intertial forces induced by the jth mode, referring to Figure 7.18, are as follows:

$$F_j = [L_j\ \beta\ (T_j)]\ W\ \Phi_j$$

$$= [(L_j,\ \beta\ (T_j)] \begin{bmatrix} W_1\Phi_{1j} \\ W_2\Phi_{2j} \\ . \\ . \\ . \\ W_i\Phi_{ij} \\ . \\ . \\ . \end{bmatrix} = \begin{bmatrix} F_{1j} \\ F_{2j} \\ . \\ . \\ . \\ F_{ij} \\ . \\ . \\ . \end{bmatrix}$$

where W_i is the gravity load of the ith floor on the frame during the earthquake.

5. Using the square root of the squares approximation of $\sqrt{(\Sigma\ F_{ij}^2)}$, and referring to Figure 7.19, the total intertial forces response on the structure is as follows:

$$F_T = \begin{bmatrix} F_1 \\ F_2 \\ . \\ . \\ . \\ F_i \\ . \\ . \\ . \end{bmatrix} = \begin{bmatrix} \sqrt{(\Sigma\ F^2_{1j})} \\ \sqrt{(\Sigma\ F^2_{2j})} \\ . \\ . \\ . \\ \sqrt{(\Sigma\ F^2_{ij})} \\ . \end{bmatrix}$$

6. The total base shear $= \Sigma\ F_i$.

7.6 TOTAL INERTIAL RESPONSE FORCES ON A FRAME

A four-storey single-bay frame is chosen for this purpose, which is defined below. All the members of the frame are taken to have a Young's modulus $E = 30 \times 10^6$ kN/m^2; the cross-sectional sizes of beams and columns are 0.5×0.6 m^2 and 0.5×0.5 m^2 respectively.

Two frame idealizations are used. One corresponds to the normal elastic plane frame, taking into account axial shortening and joint rotations. The other correspond to the 'shear building' approximation, where the axial effects are neglected and the beams' second moment of area is assumed to be infinite.

The peak ground acceleration $\alpha = 0.25$, the frame's behaviour factor $q = 3.0$, soil class A is assumed and the soil factor $S = 1.0$; the response spectrum $T_c = 0.4$ seconds and the damping of the frame $\mu = 0.05$. It is also assumed that the importance factor $I = 1.0$.

The gravity loads of the floors are taken to be equal to 400 kN with the exception of the first floor which is 800 kN.

7.6.1 First mode analysis

First mode analysis is allowed by seismic codes for mainly regular buildings, and in order to find the seismic forces acting on each storey of the building it is necessary to establish the fundamental period T seconds of the structure. There are a number of ways of approximating the period T and values can vary widely, as illustrated below.

(a) Empirical expression for T

The formula below is found in most seismic codes around the world and applies for frames:

$$T \simeq 0.1 \times (\text{number of storeys}) = 0.4 \text{ seconds}$$

(b) EC8 method

The expression for T is

$$T = 2\sqrt{\delta}$$

where δ is the horizontal deflection (in metres) of the top storey of the building under consideration when loaded laterally at each floor with forces equal to the gravity loads of the corresponding floor.

(i) Plane frame
From a plane frame analysis computer program, it was found that $\delta = 0.0856$ m, and hence the natural period is

$$T = 2\sqrt{0.0856} = 0.585 \text{ seconds}$$

(ii) Shear building
The deflection across the storey of the shear frame of height h and second moment of area IZ, when acted upon by a shear F, is given by the expression $F = 24E(IZ)\Delta/h^3$. Hence the top deflection

$$\delta = 11 \times (400 \times 3^3)/(24 \times 30\,000\,000 \times 0.5^4/12)$$

$$= 0.032 \text{ m}$$

Thus

$$T = 2\sqrt{0.032} = 0.358 \text{ seconds}$$

(c) Rayleigh's method

The expression for T is

$$T = 2\pi\sqrt{[(\Sigma W\Phi^2)/(g\Sigma W\Phi)]}$$

where W are the gravity loads at each floor and Φ their lateral displacements when the loads W are applied horizontally to the structure.

(i) Plane frame
From a plane frame analysis computer program, it was found that the fundamental mode deflection for $W = [400, 400, 400, 800]^T$ is

$$\Phi = [0.0856, 0.074, 0.0539, 0.0256]^T$$

Hence period

$$T = 0.509 \text{ seconds}$$

(ii) Shear building
By hand calculation the deflected form

$$\varphi = [0.032, 0.0291, 0.233, 0.0145]^T$$

and hence

$$T = 0.317 \text{ seconds}$$

(d) Use of response spectrum

It can be seen from the response spectrum for EC8 that when the value of the period is less than or equal to $T = 0.4$ s, the value of the response acceleration of the mass is given by:

$$S(T) = \beta_c (T) \, \alpha(S/q)$$

$$= (2.5) \times 0.25 \times (1.0/3.0) = 0.208$$

while when $T > 0.4$ s then

$$S(T) = 0.208 \times (T_c/T)^{2/3}$$

and hence the lowest coefficient corresponding to the periods found above is

$$S(T) = 0.208 \times (0.4/0.585)^{2/3}$$

$$= 0.161$$

(e) Inertial response floor forces

(i)   Shear building
Total gravity load $= 2000$ kN. Base shear $F_{BS} = 0.208 \times 2000 = 416$ kN. Thus

$$F_i = [W_i h_i / \sum(Wh)]F_{BS}$$

$$\sum(Wh) = 400 \times (2 \times 3 + 6 + 9 + 12)$$

$$= 13\,200 \text{ kN m}$$

Seismic response floor forces (kN):

$$F_1 = (400 \times 12/13\,200) \times 416 = 151$$

$$F_2 = (400 \times 9/13\,200) \times 416 = 113$$

$$F_3 = (400 \times 6/13\,200) \times 416 = 76$$

$$F_4 = (800 \times 3/13\,200) \times 416 = 76$$

which add up to a base shear of 416 kN.

(ii)   Plane frame
Using the first mode analysis computer program given below for $T = 0.585$ s, we get the seismic response floor forces (kN):

$$F_1 = 118$$

$$F_2 = 88$$

$$F_3 = 59$$

$$F_4 = 59$$

which add up to a base shear of 314 kN.

7.6.2 Modal analysis

Two cases have been investigated using the modal analysis computer program given below, for the shear building and the plane frame cases.

The programe requires one to specify the stiffness matrix and the mass matrix of the frame. The program outputs the values of Ω^2, the period T and eigenvectors for each of the modes of vibration. The participation factor for each mode is evaluated and used to obtain the seismic floor response forces for the square root of the sum of squares (SRSS) procedure (recommended in EC8 (Draft)).

(a) Stiffness matrix

(i)   Shear building

$$K = \begin{bmatrix} 138871 & -138871 & 0 & 0 \\ -138871 & 277742 & -138871 & 0 \\ 0 & -138871 & 277742 & 138871 \\ 0 & 0 & -138871 & 277742 \end{bmatrix}$$

The above coefficients have been obtained by long-hand calculation using the well-known expressions such as $K_{11} = 24E(IZ)/h^3$ and $K_{22} = 48E(IZ)/h^3$, etc.

(ii)   Plane frame

$$K = \begin{bmatrix} 68854 & -89858 & 23009 & -2164 \\ -89858 & 189264 & -123112 & 27355 \\ 23009 & -123112 & 201146 & -128905 \\ -2164 & 27355 & -128905 & 232508 \end{bmatrix}$$

The above stiffness matrix has been obtained by using a plane frame computer program. The approach used was first to impose horizontal restraints at all floors, and then allow each one in turn to move unit displacement; i.e. four analyses have been carried out and from each one the corresponding column of the above stiffness matrix has been obtained.

(b) Mass matrix

The mass matrix is common to both idealizations and is given as below:

$$M = \begin{bmatrix} 40.775 & 0 & 0 & 0 \\ 0 & 40.775 & 0 & 0 \\ 0 & 0 & 40.775 & 0 \\ 0 & 0 & 0 & 81.55 \end{bmatrix}$$

(c) Results of computer run

(i) Shear building
The results in this case are shown in Table 7.2. Seismic response floor forces (kN):

$$F_1 = 115$$

$$F_2 = 98$$

$$F_3 = 77$$

$$F_4 = 111$$

giving a base shear equal to 401 kN.

Table 7.2 Results of computer run for shear building

Period(s) $T_1 = 0.318$	$T_2 = 0.127$	$T_3 = 0.079$	$T_4 = 0.059$
Modes of			
vibration 0.098	−0.08	−0.08	−0.047
0.087	−0.023	0.067	0.108
0.066	0.050	0.089	−0.098
0.037	0.087	−0.054	0.021
Participation			
factors 27.233	10.19	−2.402	0.526

Table 7.3 Results of computer run for plane frame building

Period(s) $T_1 = 0.515$	$T_2 = 0.177$	$T_3 = 0.102$	$T_4 = 0.066$
Modes of			
vibration 0.106	−0.083	−0.07	−0.039
0.089	0.001	0.082	0.098
0.061	0.078	0.058	−0.106
0.025	0.075	−0.07	0.032
Participation			
factors 30.58	12.412	−5.512	1.723

(ii) Plane frame
The results in this case are shown in Table 7.3. Seismic response floor forces (kN):

$$F_1 = 104$$

$$F_2 = 82$$

$$F_3 = 69$$

$$F_4 = 94$$

giving a base shear equal to 349 kN.

It can be seen that the periods are sensitive to the structural idealization, giving somewhat differing response forces; it is advisable to take the shear building case periods, as in real buildings partitions and other fixtures will have the effect of reducing the periods of vibration of the structure. In addition, the above assumption is on the safe side as the response acceleration will tend to be higher.

7.6.3 First mode computer seismic response floor forces

See computer program 5.

7.6.4 Computer program for the evaluation of Ω, *T*, modes of vibration, participation factors and seismic force distribution (SRSS)

See computer program 6.

REFERENCES AND BIBLIOGRAPHY

Ambraseys N.N. and Jackson J.A. (1981) Earthquake hazard and vulnerability in the north eastern Mediterranean: Corinth earthquake sequence of February–March 1981, *Disasters*, **5**, No. 4, 355–68.

Anderson J.A. and Wood H.O. (1925) Description and theory of the torsion tensometer, *Bull. Seism. Soc. Am.*, **15**, 1.

Arnold C. (1984) Soft first storeys: truths and myths, *Proc. 8th World Conf. on Earthq. Eng.*, San Francisco, vol. V, pp. 943–9.

Dowrick D.J. (1987) *Earthquake Resistant Design, for Engineers and Architects*, Wiley, New York.

Gutenberg B. and Richter C.F. (1965) *Seismicity of the Earth*, Stechert-Hafner, New York.

Hanks T.C. and Kanamori H. (1979) A moment magnitude scale, *J. Geophys. Res.*, **84**, B5, 2348–50.

Hodgson J.H. (1964) *Earthquakes and Earth Structure*, Prentice-Hall, Englewood Cliffs, NJ.

```
      CLS
      REM Floor gravity loads
      DATA 400, 400, 400, 800
      REM Floor heights above the foundation
      DATA 12, 9, 6, 3
      INPUT "NUMBER OF STOREYS N = "; N

      DIM FORCE(N), FW(N), FH(N)
      FOR I = 1 TO N: READ FW(I): NEXT
      FOR I = 1 TO N: READ FH(I): NEXT

      REM Response spectrum
7020  INPUT "PEAK GROUND ACCELERATION RATIO = "; ALPHA
      IF ALPHA < 0 OR ALPHA > 0.5 THEN 7020
7220  INPUT "SOIL TYPE A, B OR C = "; SOIL$
      IF SOIL$ < CHR$(64) OR SOIL$ > CHR$(67) THEN 7220
7420  INPUT "DUCTILITY BEHAVIOUR FACTOR q = "; Q: PRINT
      IF Q < 1 OR Q > 4 THEN 7420
7520  INPUT "PERIOD OF VIBRATION IN SECONDS T = "; T: PRINT
      IF T < (0.05 * N) OR T > (0.15 * N) THEN PRINT "IS IT SO? Y/N"
      INPUT A$: IF A$ = "N" THEN 7520

      FOR I = 1 TO N
      IF SOIL$ = "A" THEN T2 = 0.4 : SOL = 1
      IF SOIL$ = "B" THEN T2 = 0.6 : SOL = 1
      IF SOIL$ = "C" THEN T2 = 0.8 : SOL = 0.9
      Z = 2.5 * ALPHA * SOL/Q
      Z1 = Z * (T2/T)^ 0.667
      Z2 = 0.2 * ALPHA
      IF T < = T2 THEN BETA = Z ELSE BETA = Z1
      IF BETA < Z2 THEN BETA = Z2
      NEXT

      REM Seismic response floor forces
      SW = 0
      FOR I = 1 TO N: SW = SW + FW(I): NEXT
      SWH = 0
      FOR I = 1 TO N: SWH = SWH + FW(I) * FH(I): NEXT
      FBS = BETA * SW
      FOR I = 1 TO N: FK = FH(I) * FW(I)/SWH
      FORCE(I) = FK * FBS
      NEXT
      PRINT "SEISMIC FORCES": PRINT
      FOR I = 1 TO N
      PRINT INT(FORCE(I) * 1000)/1000: PRINT
      NEXT
      END
```

Computer program 5

```
10      NN = 30

        DIM KK(NN, NN), K#(NN, NN), MM(NN, NN), M#(NN, NN), X#(NN, NN)
        DIM eigv#(NN), D#(NN), T(NN), p(NN)
        DIM BETA(NN), F(NN, NN), FORCE(NN)

        REM STRUCTURE STIFFNESS

        DATA 569984, -568800, 0, 0
        DATA -568800, 1138352, -568800, 0
        DATA 0, -568800, 1138352, -568800
        DATA 0, 0, -568800, 1138352

        REM STRUCTURE MASS MATRIX

        DATA 203.9, 0, 0, 0
        DATA 0, 101.9, 0, 0
        DATA 0, 0, 101.9, 0
        DATA 0, 0, 0, 101.9

        N = 4

        FOR I = 1 TO N: FOR J = 1 TO N: READ K#(I, J): KK(I, J) = K#(I, J)
        NEXT J, I
        FOR I = 1 TO N: FOR J = 1 TO N: PRINT K#(I, J): NEXT J: PRINT NEXT I
        FOR I = 1 TO N: FOR J = 1 TO N: READ M#(I, J): MM(I, J) = M#(I, J)
        NEXT J, I
        FOR I = 1 TO N: FOR J = 1 TO N: PRINT M#(I, J): NEXT J: PRINT: NEXT

        REM Jacobi's eigenvalues and mass normalized vectors, solution

        NSMAX = 20: RTOL = 1E-11: IFPR = 0
        FOR I = 1 TO N
        IF K#(I, I) > 0 AND M#(I, I) > 0 THEN 599
        PRINT "ERROR, MATRICES NOT POSITIVE DEFINITE"
        END

599     D#(I) = K#(I, I)/M#(I, I)
        eigv#(I) = D#(I)
        NEXT I
        FOR I = 1 TO N: FOR J = 1 TO N: X#(I, J) = 0: NEXT J
        X#(I, I) = 1: NEXT I
799     IF N = 1 THEN 2199
        NSWEEP = 0
        NR = N - 1
899     NSWEEP = NSWEEP + 1
        IF IFPR = 1 THEN PRINT "SWEEP NUMBER IN JACOBI = "; NSWEEP
        EPS# = (0.01^NSWEEP)^2
        FOR J = 1 TO NR
        JJ = J + 1
        FOR K = JJ TO N
        EPTOLA# = K#(J, K) * K#(J, K)/(K#(J, J) * K#(K, K))
        EPTOLB# = (M#(J, K) * M#(J, K))/(M#(J, J) * M#(K, K))
980     IF EPTOLA# < EPS# AND EPTOLB# < EPS# THEN 1690
        AKK# = K#(K, K) * M#(J, K) - M#(K, K) * K#(J, K)
        AJJ# = K#(J, J) * M#(J, K) - M#(J, J) * K#(J, K)
        AB# = K#(J, J) * M#(K, K) - K#(K, K) * M#(J, J)
        CHECK# = (AB# * AB# + 4# * AKK# * AJJ#)/4#
1030    IF CHECK# >= 0 THEN 1050
        PRINT "ERROR, MATRICES NOT POSITIVE DEFINITE"
        END
```

Computer program 6

```
1050  SQCH# = SQR(CHECK#)
      D1# = AB#/2# + SQCH#
      D2# = AB#/2# - SQCH#
      DEN# = D1#
      IF ABS(D2#) > ABS(D1#) THEN DEN# = D2#
1080  IF DEN# <> 0 THEN 1100
1090  CA# = 0: CG# = -K#(J, K)/K#(K, K): GOTO 1200
1100  CA# = AKK#/DEN#
      CG# = -AJJ#/DEN#
1200  IF N - 2 = 0 THEN 1499
      JP1 = J + 1: JM1 = J - 1: KP1 = K + 1: KM1 = K - 1
1220  IF JM1 - 1 < 0 THEN 1299
      FOR I = 1 TO JM1
      AJ# = K#(I, J): BJ# = M#(I, J)
      AK# = K#(I, K): BK# = M#(I, K)
      K#(I, J) = AJ# + CG# * AK#
      M#(I, J) = BJ# + CG# * BK#
      K#(I, K) = AK# + CA# * AJ#
      M#(I, K) = BK# + CA# * BJ#
      NEXT I
1299  IF KP1 - N > 0 THEN 1399
      FOR I = KP1 TO N
      AJ# = K#(J, I): BJ# = M#(J, I)
      AK# = K#(K, I): BK# = M#(K, I)
      K#(J, I) = AJ# + CG# * AK#
      M#(J, I) = BJ# + CG# * BK#
      K#(K, I) = AK# + CA# * AJ#
      M#(K, I) = BK# + CA# * BJ#
      NEXT I
1399  IF JP1 - KM1 > 0 THEN 1499
      FOR I = JP1 TO KM1
      AJ# = K#(J, I): BJ# = M#(J, I)
      AK# = K#(I, K): BK# = M#(I, K)
      K#(J, I) = AJ# + CG# * AK#
      M#(J, I) = BJ# + CG# * BK#
      K#(I, K) = AK# + CA# * AJ#
      M#(I, K) = BK# + CA# * BJ#
      NEXT I
1499  AK# = K#(K, K): BK# = M#(K, K)
      K#(K, K) = AK# + 2# * CA# * K#(J, K) + CA# * CA# * K#(J, J)
      M#(K, K) = BK# + 2# * CA# * M#(J, K) + CA# * CA# * M#(J, J)
      K#(J, J) = K#(J, J) + 2# * CG# * K#(J, K) + CG# * CG# * AK#
      M#(J, J) = M#(J, J) + 2# * CG# * M#(J, K) + CG# * CG# * BK#
      K#(J, K) = 0: M#(J, K) = 0
      FOR I = 1 TO N
      XJ# = X#(I, J): XK# = X#(I, K)
      X#(I, J) = XJ# + CG# * XK#
      X#(I, K) = XK# + CA# * XJ#
      NEXT I
1690 NEXT K
      NEXT J
      FOR I = 1 TO N
1760  IF K#(I, I) > 0 AND M#(I, I) > 0 THEN 1790
      PRINT "ERROR, MATRIX NOT POSITIVE DEFINITE"
      END
1790  eigv#(I) = K#(I, I)/M#(I, I)
      NEXT I
1810  IF IFPR = 0 THEN 1899
```

Computer program 6 *continued*

```
          PRINT "CURRENT EIGENVALUES ARE"
1899   FOR I = 1 TO N
          TOL = RTOL * D(I)
          DIF = ABS(eigv#(I) - D#(I))
1930   IF DIF > TOL THEN 2210
          NEXT I
          EPS = RTOL^2
          FOR J = 1 TO NR
          JJ = J + 1
          FOR K = JJ TO N
          EPSA# = K#(J, K) * K#(J, K)/(K#(J, J) * K#(K, K))
          EPSB# = M#(J, K) * M#(J, K)/(M#(J, J) * M#(K, K))
2010   IF EPSA# < EPS# AND EPSB# < EPS# THEN 2090
2080   GOTO 2210
2090   NEXT K
          NEXT J
2105   PRINT
          FOR I = 1 TO N
          FOR J = 1 TO N
          K#(J, I) = K#(I, J)
          M#(J, I) = M#(I, J)
          NEXT J
          NEXT I
          FOR J = 1 TO N
          BB# = SQR(M#(J, J))
          FOR K = 1 TO N
          X#(K, J) = X#(K, J)/BB#
          NEXT K
          NEXT J
2194   GOTO 4000
2195   PRINT
2199   END

2210   FOR I = 1 TO N: D#(I) = eigv#(I)
          NEXT I
2230   IF NSWEEP < NSMAX THEN 899
2240   GOTO 2105

4000   REM placing eigenvalues in ascending order

          FOR I = 1 TO N - 1: K = I: p# = eigv#(I)
          FOR J = I + 1 TO N
          IF eigv#(J) >= p# THEN 4050
          K = J: p# = eigv#(J)
4050   PRINT
          NEXT J
          IF K = I THEN 4110
          eigv#(K) = eigv#(I): eigv#(I) = p#

          REM reorganizing eigenvectors

          FOR J = 1 TO N
          p# = X#(J, I): X#(J, I) = X#(J, K): X#(J, K) = p#
          NEXT J
4110   PRINT
          NEXT I

          REM evaluation of the periods

          FOR I = 1 TO N: F = SQR(eigv#(I)): T(I) = 2 * 3.1414/F: NEXT
```

Computer program 6 *continued*

```
      REM participation factors

      FOR I = 1 TO N: C = 0: FOR J = 1 TO N: C = C + MM(I, J)
      NEXT: MIHAT(I) = C: NEXT
      FOR I = 1 TO N: p = 0: FOR J = 1 TO N
      p = p + X#(J, I) * MIHAT(J): NEXT
      p(I) = p
      NEXT
CLS

      PRINT: PRINT
      PRINT "******INPUT DATA******"
      PRINT

      PRINT "STRUCTURE STIFFNESS"
      PRINT
      FOR I = 1 TO N: FOR J = 1 TO N: K = KK(I, J): PRINT K: NEXT PRINT
      NEXT
      PRINT

      PRINT "STRUCTURE MASS"
      PRINT
      FOR I = 1 TO N: FOR J = 1 TO N: M = MM(I, J): PRINT M: NEXT PRINT
      NEXT
      PRINT
4999  INPUT "NUMBER OF EFFECTIVE MODES = "; NE

      IF NE <= 0 THEN 4999
      IF NE > N THEN NE = N
      PRINT

7020  INPUT "PEAK GROUND ACCELERATION = "; ALPHA
      IF ALPHA < 0 OR ALPHA > 0.5 THEN 7020
7220  INPUT "SOIL TYPE A, B OR C = "; SOIL$
      IF SOIL$ < CHR$ (64) OR SOIL$ > CHR$ (67) THEN 7220
7420  INPUT "DUCTILITY BEHAVIOUR FACTOR q = "; Q
      IF Q < 1 OR Q > 5 THEN 7420

      GG = 9.81
      MU = 0.05
      FOR I = 1 TO N
      IF SOIL$ = "A" THEN T2 = 0.4: SOL = 1
      IF SOIL$ = "B" THEN T2 = 0.6: SOL = 1
      IF SOIL$ = "C" THEN T2 = 0.8: SOL = 0.8
      Z = 2.5 * ALPHA * GG * SQR(0.05/MU) * SOL/Q
      Z1 = Z * (T2/T(I))^0.667
      Z2 = 0.2 * ALPHA * GG
      IF T(I) <= T2 THEN BETA(I) = Z ELSE BETA(I) = Z1
      IF BETA(I) < Z2 THEN BETA(I) = Z2
      NEXT

      FOR I = 1 TO N: p(I) = p(I) * BETA(I): NEXT
      FOR J = 1 TO N
      FOR I = 1 TO N
      F(I, J) = MM(I, I) * X#(I, J) * p(J)
      NEXT I
      NEXT J
```

Computer program 6 *continued*

```
BSH = 0
FOR I = 1 TO N
SUM = 0
FOR J = 1 TO N
SUM = SUM + F(I, J)^2
NEXT J
FORCE(I) = SQR(SUM): BSH = BSH + FORCE(I)
NEXT I

PRINT
PRINT "OMEGA SQUARE"
PRINT
FOR I = 1 TO NE
PRINT INT(eigv#(I) * 1000)/1000
NEXT
PRINT
PRINT
PRINT "PERIOD IN SECONDS"
PRINT
FOR I = 1 TO NE
PRINT INT(T(I) * 1000)/1000
NEXT

PRINT
PRINT
PRINT "EIGENVECTORS"
PRINT
FOR I = 1 TO N: FOR J = 1 TO NE: X = X#(I, J)
PRINT INT(X * 1000)/1000: NEXT: PRINT: NEXT
PRINT
PRINT "PARTICIPATION FACTORS"
PRINT
FOR I = 1 TO NE: PRINT INT(p(I) * 1000)/1000: NEXT: PRINT
PRINT
PRINT "SRSS FORCES"
PRINT
FOR I = 1 TO N
PRINT INT(FORCE(I) * 1000)/1000
NEXT
GOTO 2195
```

Computer program 6 *continued*

Karnik V. (1969, 1971) *Seismicity of the European Area*, parts 1 and 2, Reidel, Dordrecht.

Kasamara, K. (1980) *Earthquake Mechanics*, Cambridge University Press, Cambridge.

Key D. (1988) *Earthquake Design Practice for Buildings*, Thomas Telford, London.

Paz M. (1980) *Structural Dynamics Theory and Computations*, Van Nostrand Reinhold, New York.

Penzien J. (1969) Earthquake response of irregularly shaped buildings, *Proc. 4th World Conf. on Earthq. Eng.*, Santiago, vol. 2, A3, pp. 75–90.

Redmayne D. (1993) Recent notable earthquakes 1980–1993, *SECED Newsletter*, **7**, No. 4, pp. 14–17.

Richter C.F. (1935) An instrumental earthquake magnitude scale, *Bull. Seism. Soc. Am.*, **25**, 1.

Scruton C. (1963) *Wind Effects on Buildings and Structures*, vol. II, HMSO, London, pp. 797–832.

Seed H.B. (1969) The influence of local soil conditions on earthquake damage, *Proc. 7th Int. Conf. on Soil Mech. Found. Eng.*, Mexico, Specialty Session 2, pp. 33–66.

Seed H.B., Ugas C. and Lysmer J. (1974) *Site Dependent Spectra for Earthquake Resistant Design*, Report No. EERC 74–12, Earthquake Engineering Centre, University of California, Berkeley.

Wiegel R.L. (ed.) (1970) *Earthquake Engineering*, Prentice-Hall, Englewood Cliffs, NJ.

Wood H.O. and Neumann F. (1967) Modified Mercalli intensity scale of 1931, *Bull. Seism. Soc. Am.*, **23**, 277.

8

DESIGN EXAMPLE

8.1 INTRODUCTION

Chapters 4, 5 and 6 included a number of examples of element design (slabs, beams and columns) for both the ultimate and serviceability limit states. In this chapter, the design of a six-storey office building is presented and two cases are considered:

(i) Design to EC2/NAD (permanent, variable and wind actions)
(ii) Design to EC8/EC2 (permanent, variable and seismic actions).

The general arrangement of a typical floor is shown in Figure 8.1; the storey heights are 3.5 m with the exception of the ground to first floor, which is 5.0 m. For the design to EC2/NAD, the building is assumed to be located on the outskirts of Oxford. In the case of the design to EC8/EC2, it is assumed that the building is located in southern Europe where there is a record of seismic activity. Referring to Figure 8.1, the following will be considered for preliminary design – the end bay of the floor slab, the line of continuous beams ABCD and the six-storey frame associated with beam line ABCD. The variable action Q_k is taken as constant for all the floors and the roof (this allows for possible future extension). The basic design data are given in Table 8.1.

Prior to undertaking a detailed design involving a computer analysis (not included), a preliminary design is necessary to check that the trial element dimensions are appropriate and that the reinforcement ratio limits are not exceeded. The preliminary design is covered in sections 8.2 and 8.3.

8.2 PRELIMINARY DESIGN: CASE (i) EC2/NAD

8.2.1 Loading – general

The building is located on the outskirts of Oxford and thus, from CP3: Chapter V: Part 2, the basic wind speed is 40 m/s. Consider wind acting on the longer (30 m) face of the building. Take $S_1 = S_3 = 1.0$ and S_2 is obtained from table 3 of CP3: Part 2, as below (assuming class B, exposure condition (1)):

$$S_3 = 1.01 + (1.05 - 1.01) \times 0.25 = 1.02$$

This gives the design wind speed

$$V_s = 1.02 \times 40 = 40.8 \text{ m/s}$$

and dynamic pressure

$$q = 0.613 \times 40.8^2 = 1020.4 \text{ N/m}^2$$

From table 10 of CP3: Part 2, the force coefficient C_f approximates to 0.97 and thus wind pressure is

$$p = 0.97 \times 1020.4 \times 10^{-3} = 0.99 \quad \text{(say 1.0 kN/m}^2\text{)}$$

Conservatively, this is assumed to be constant over the full height of the building and the wind forces (unfactored) resisted by frame ABCD at floor and roof levels are:

roof $1.0 \times 1.75 \times 5 = 8.75$ kN

2nd to 5th floors $1.0 \times 3.5 \times 5 = 17.5$ kN

DESIGN EXAMPLE

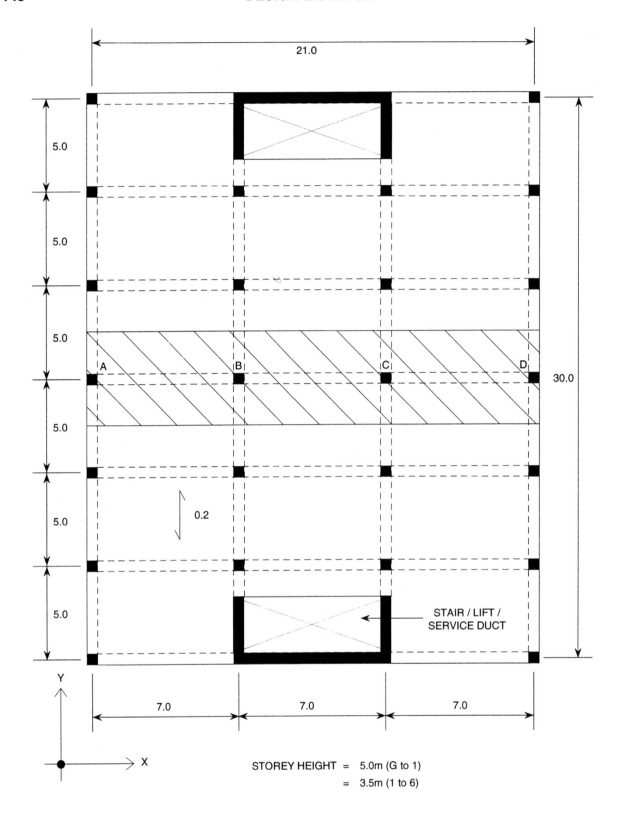

Figure 8.1 Typical floor plan assumed for preliminary design of frame.

Table 8.1   Basic design data

	EC2/NAD	EC8/EC2	Comment
Char. cyl. str., f_{ck} (N/mm²)	30	30	
Char. steel str., f_{yk} (N/mm²)	460	400	400 N/mm² more appropriate for seismic actions
Unit wt. of concrete (kN/m³)	24	24	
Partial safety factors (ULS)			
γ_G	1.35	–	See section 8.3.1
γ_Q	1.5	–	See section 8.3.1
γ_c	1.5	1.5	
γ_s	1.15	1.15	
Variable action, Q_k (kN/m²)	5.0	5.0	Client requirement for future adaption, exceeds Code requirements for offices
Exposure class	2b	2a	2b (frost, UK) 2a (no frost, S. Europe)
Basic wind speed (m/s)	40	–	CP3: Part 2, class B, cat. (1)
Seismic coefficient, DC 'M'	–	0.15	See section 8.3.1
Reinforcement ratio (flexure)			
minimum	0.0015	0.0036	See equation, (4.20) and Table 4.8
maximum	0.015	0.026	See Tables 4.1 and 4.8
Basic shear str., τ_{Rd} (N/mm²)	0.34	–	See Table 1.6
Mean tensile str., f_{ctm} (N/mm²)	2.9	2.9	See Table 1.6
Mod. of elasticity, E_{cm} (kN/mm²)	32	32	See Table 1.6

1st floor $1.0 \times 4.25 \times 5 = 21.25$ kN

The trial depth for the slab is 0.2 m, say, thus

self-weight $= 0.2 \times 24 = 4.8$ kN/m²

Allow 2.2 kN/m² for partitions, services, suspended ceiling and finishes. Thus:

$G_k = 4.8 + 2.2 = 7.0$ kN/m²

and

$Q_k = 5.0$ kN/m²

The numbering of frame ABCD, the wind loadings and trial slab, beam and column dimensions are shown in Figure 8.2. For simplicity, the following loading arrangements will be considered:

- permanent + variable, $(1.35G_k + 1.5Q_k)$, on all spans for slabs and beams
- permanent + variable + wind, $1.35 (G_k + Q_k +$ wind).

The error involved in not considering pattern loading for Q_k is (in this design) small.

8.2.2 Slab (ULS) – end bay

The aspect ratio of the slab (7.0/5.0 = 1.4) is such that, in a detailed design, two-way spanning should

be considered using elastic bending moment coefficients (see BS 8110: 1985 or Timoshenko and Woinowsky-Krieger (1959)) or plastic analysis (yield line or the strip method; see Appendices D and E). For preliminary design, it is adequate to assume that the slab is one-way continuous and to use the approximate bending moment coefficients for a four-span slab; see Figure 3.3, case 8. At the penultimate support, the design bending moment is:

$M_d = 0.107 \ (1.35 \ G_k + 1.5 \ Q_k) \ L^2$

For a 1 m width

$1.35G_k = 1.35 \times 7.0 = 9.45$ kN/m

and

$1.5 \ Q_k = 1.5 \times 5.0 = 7.50$ kN/m

Thus

$M_d = 0.107 \times 16.95 \times 5^2 = 45.34$ kN m

For exposure class 2b, the NAD (see Table 2.8) gives a cover of 35–5 = 30 mm for slabs. Assuming 12 mm dia. bars, the effective depth is:

$d = 200 - (30 + 6) = 164$ mm

From the design chart for flexure, see Figure 4.3:

$M_d/b_w d^2 = 45.34 \times 10^6/10^3 \times 164^2 = 1.69$

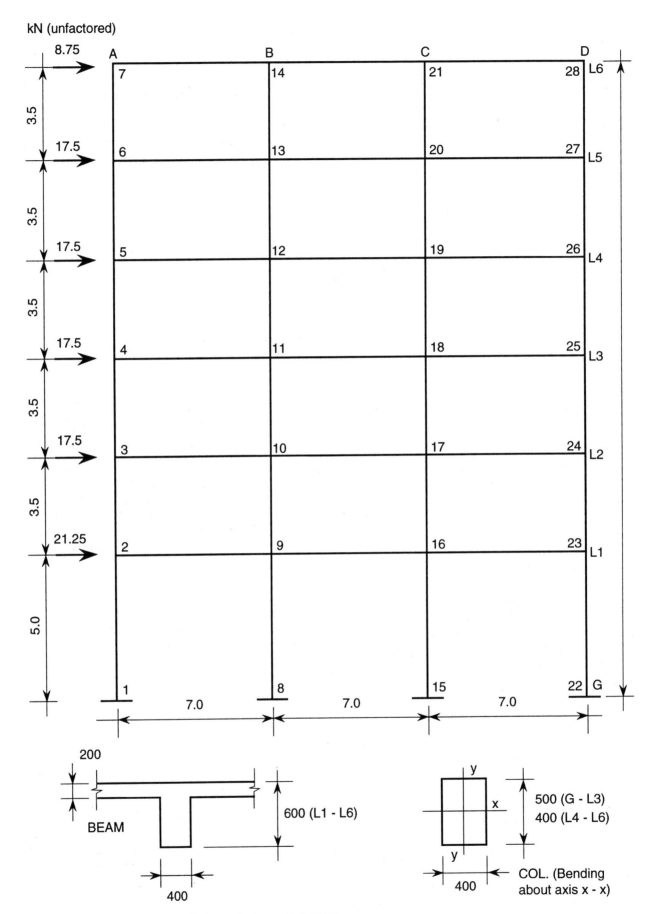

kN (unfactored)

Figure 8.2 Frame ABCD – preliminary design to EC2/NAD.

Thus for $f_{ck} = 30$ N/mm²:

$x/d = 0.13$

$d = 21.32$ mm

$A_s = 45.34 \times 10^6/(460/1.15) \times (164 - 0.4 \times 21.32)$

$\quad = 729$ mm²/m ($12\varphi - 150 = 754$)

Then

reinforcement ratio $= 754/10^3 \times 164$

$\qquad\qquad\qquad\quad = 0.0046$ (0.46%)

The design shear force at the penultimate support is

$V_d = 16.95 \times 2.5 + 45.34/5$

$\quad = 51.44$ kN

The shear stress

$v = 51.44 \times 10^3/10^3 \times 164 = 0.31$ N/mm²

From Table 8.1 the basic shear strength is $\tau_{Rd} = 0.34$ N/mm²; thus the slab is adequate in shear without considering enhancement factors for depth and reinforcement ratio (see equation (4.2)).

8.2.3 Slab (SLS) – end bay

The slab is lightly stressed, steel percentage $p < 0.5$ and the span/effective depth ratio is $5/0.164 = 30.49$. As f_{yk} is 460 N/mm², a rapid check on the steel stress σ_s at service load will be made. At service load $G_k + Q_k = 12$ kN/m and putting the combination factor $\psi = 1.0$ to compensate for the underestimation of the bending moment in the span as pattern loading is not considered, then

$M_{(span)} = 0.077 \times 12 \times 5^2 = 23.1$ kN m

The modular ratio $\alpha_e = 200/32 = 6.25$ and thus if the same reinforcement percentage is provided in the span (0.46), the neutral axis factor ($n \sim 0.23$) can be obtained from Figure 4.19. The lever arm is $164(1 - n/3) = 151.43$ mm. Thus the steel stress σ_s at service load approximates to

$\sigma_s = 23.1 \times 10^6/754 \times 151.43 = 202.3$ N/mm²

As σ_s is less than 250 N/mm² (see EC2 cl. 4.4.3.2), the basic L/d ratio of 32 for the end span of a one-way continuous slab or two-way spanning slab continuous over one longer side (case 2 in table 4.14 of EC2) does not require any modification. Thus the L/d ratio of 30.49 is satisfactory. For cracks caused dominantly by loading, the provisions of table 4.11 or table 4.12 of EC2 must be complied with. The steel stress used should be evaluated for the quasi-permanent loads. From table 1 of the NAD, the combination factor $\psi_2 = 0.3$ for offices. The bending moment at the penultimate support is

$M_{SLS} = 0.107 (7 + 0.3 \times 5) \times 5^2 = 22.74$ kN m

By inspection, the stress level associated with the above moment (see deflection calculations, Chapter 4) will satisfy the requirements of tables 4.11 and 4.12 of EC2.

8.2.4 Beam line ABCD (ULS)

(a) Permanent + variable loads, $1.35\,G_k + 1.5\,Q_k$

The loading of the beam from the slab is $16.95 \times 5 = 34.75$ kN/m, to which must be added the self-weight of the beam rib. From Figure 8.2 this is $0.4 \times 0.4 \times 24 = 3.84$ kN/m. This must be multiplied by the partial safety factor $\gamma_f = 1.35$ for permanent loading, giving a total load of $84.75 + 3.84 \times 1.35 = 89.934$ kN/m. The preliminary beam design will concentrate on level 1 (beam line 2–9–16–23); see Figure 8.2. From Figure 3.3, load case 3, the maximum beam moment at joints 9 and 16 approximates to:

$M_d = 0.1 \times 89.934 \times 7^2 = 440.68$ kN m

Assuming simple end supports at 2 and 23, the maximum beam shear force at joint 9 is:

$V_d = 89.934 \times 3.5 + 440.68/7$

$\quad = 314.77 + 62.95$

$\quad = 377.72$ kN

(b) Permanent + variable + wind loads, $1.35(G_k + Q_k + wind)$

For this case, the floor loading (ULS) is $1.35 (7 + 5) = 16.2$ kN/m. Thus the load per metre run of beam is $16.2 \times 5 + 3.84 \times 1.35 = 86.184$ kN. Thus the maximum bending moment at joint 9 is:

$M_d = 440.68 \times 86.184/89.934$

$\quad = 422.3$ kN m

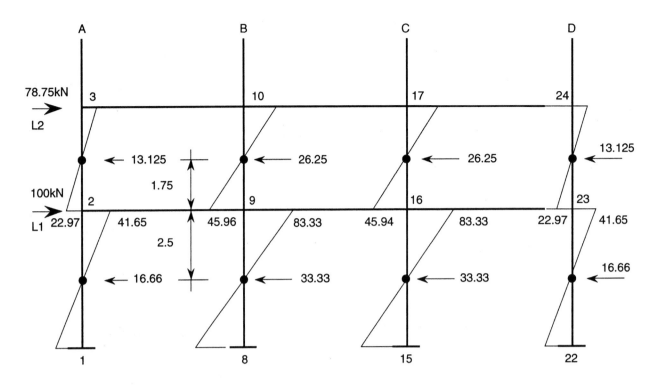

(a) Column Moments

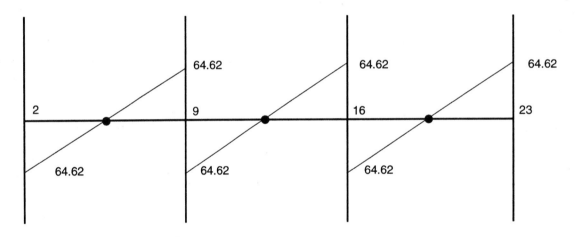

(b) Beam Moments

Total wind shear at L2   =   8.75   +   4 x 17.5   =   78.75kN

Total wind shear at L1   =   78.75   +   21.25   =   100.0kN

Wind loads unfactored

Figure 8.3   Wind analysis for beam and column moments.

The maximum beam shear force at joint 9 is:

$$V_d = 86.184 \times 3.5 + 422.3/7$$

$$= 301.64 + 60.33$$

$$= 361.97 \text{ kN}$$

Wind load moments and shears will be estimated using the 'portal' method as outlined in Chapter 3. Referring to Figure 8.3, the beam moment at joint 9 due to wind action is 64.62 kN and this is multiplied by a partial safety factor of 1.35 to give the total beam moment at joint 9 of:

$$M_d = 422.3 + 64.62 \times 1.35$$

$$= 422.3 + 87.24$$

$$= 509.54 \text{ kN m}$$

The beam shear at joint 9 due to wind action is $64.62 \times 2/7 = 18.46$ kN. Thus the total beam shear at joint 9 is:

$$V_d = 361.97 + 18.46 = 380.43 \text{ kN}$$

The section analysis (ULS) will be based on the design bending moment and shear forces for this loading case. For exposure class 2b, the beam cover is taken as 35 mm with an additional 25 mm to the centroid of the steel. So

$$d = 600 - 60 = 540 \text{ mm}$$

From Figure 4.3, with $f_{ck} = 30$ N/mm^2:

$$M_d/b_w d^2 = 509.54 \times 10^6/400 \times 540^2 = 4.37$$

Thus

$$x/d = 0.39$$

$$x = 210.6 \text{ mm}$$

$$d - 0.4\,x = 455.76 \text{ mm}$$

and

$$A_s = 509.54 \times 10^6/(460/1.15) \times 455.76$$

$$= 2795 \text{ mm}^2 \quad (6 - 25\varphi = 2950 \text{ mm}^2)$$

Then

$$\text{reinforcement ratio} = 2950/400 \times 540 = 0.0136$$

This is satisfactory; see Table 4.1.

The shear resistance is

$$V_{Rd1} = \tau_{Rd}\, k\, (1.2 + 40\, p_1)\, b_w d$$

Taking $\tau_{Rd} = 0.34$ N/mm^2, $k = 1.6 - 0.54 = 1.06$, then:

$$V_{Rd1} = 0.34 \times 1.06\,(1.2 + 40 \times 0.0136)$$

$$\times\, 400 \times 540 \times 10^{-3}$$

$$= 135.76 \text{ kN}$$

$V_{Rd1} < V_d$, thus shear reinforcement is required, but check that V_{Rd2} does not exceed $4.95 b_w d$ (see Table 4.2).

$$V_{Rd2} = 4.95 \times 400 \times 540 \times 10^{-3}$$

$$= 1069.2 > V_d$$

$$V_{wd} = 380.43 - 135.76 = 244.67 \text{ kN}$$

From equation (4.2), with $f_{ywd} = 460/1.15 = 400$ N/mm^2 and 8 mm φ links (four legs = 201 mm^2), the link spacing is:

$$s = 201 \times 0.9 \times 540 \times 400/244.67 \times 10^3$$

$$= 159.7 \quad \text{(say 150 mm c/c)}$$

Thus the design is adequate for flexure and shear with reasonable reinforcement requirements.

8.2.5 Beam line ABCD (SLS)

The span/effective depth ratio of the beams is $7000/540 = 12.96$ and thus the requirements of table 4.14 of EC2 are easily satisfied.

8.2.6 Columns (ULS)

(a) Permanent + variable loads, $1.35G_k + 1.5Q_k$

The load per metre run of beam is 89.934 kN and from Figure 3.3, case 3, the column reactions are:

internal $1.1 \times 89.934 \times 7 = 692.49$ kN

external $0.4 \times 89.934 \times 7 = 251.82$ kN

Making no allowance for the reduction of variable loading on the columns, the axial load on the columns between ground and first floor is:

external (1–2 and 22–23)

$$251.82 \times 6 = 1510.92 \text{ kN}$$

internal (8–9 and 15–16)

$692.49 \times 6 = 4154.94$ kN

An estimation of the column moments arising from the vertical loads may be obtained by assuming a point of contraflexure at $0.1L_B$ (see section 3.4.3(a) and Figure 3.5),

$$M_A = 0.045 \ WL_B{}^2$$

where $W = 89.934$ kN/m and $L_B = 7$ m. Thus

$$M_B = 0.045 \times 89.934 \times 7^2 = 198.3 \text{ kN m}$$

Assuming a T section for beam 2–9 with $b_{eff} = b_w + 1/5 \ l_0$ (where $l_0 = 0.85L_1$), see Figure 3.2, then:

$$b_{eff} = 0.4 + 0.2 \times 0.85 \times 7$$

$$= 1.19 + 0.4 = 1.59 \text{ m}$$

I_B approximates to 0.0128 m^4 and $I_{col} = 0.00417 \text{ m}^4$. Thus

beam stiffness	$= 0.0128/7$
	$= 0.001 \ 83 \quad (2.19)$
column stiffness (1–2)	$= 0.004 \ 17/5$
	$= 0.000 \ 834 \quad (1)$
column stiffness (2–3)	$= 0.004 \ 17/3.5$
	$= 0.001 \ 19 \quad (1.43)$

Thus moment in column 1–2 is $198.3 \times 1/4.62 = 42.92$ kN m, and moment in column 2–3 is $= 198.3 \times 1.43/4.62 = 61.38$ kN m. Thus the column moment is small compared with the axial load.

(b) Permanent + variable + wind loads, $1.35(G_k + Q_k + wind)$

The column moments arising from the vertical load of 86.184 kN/m are similar to those for the previous case. The moment in column 1–2 is $42.92 \times 86.184/89.934 = 41.13$ kN m and moment in column 2–3 is $61.38 \times 86.184/89.934 = 58.82$ kN m. To these moments the wind actions should be added, and thus from Figure 8.3 the moment in column 1–2 approximates to $41.13 + 41.65 \times 1.35 = 97.36$ kN m. The moments in the internal columns arising from vertical loading will be of small order. Thus it can be seen that axial loading dominates. From Appendix B, the ultimate axial load capacity of a column is given by:

$$N_{ud} = A_c \ (0.57 \ f_{ck} + 0.87 \ p \ f_{yk})$$

With $A_c = 200 \times 10^3 \text{ mm}^2$ and $p = 0.02$, say, then

$$N_{ud} = 200 \ (0.57 \times 30 + 0.87 \times 0.02 \times 460)$$

$$= 3420 + 1600.8$$

$$= 5020.8 \text{ kN}$$

The axial load in the column arising from wind action is obtained by taking taking moments about the point of contraflexure in the ground to first floor columns. Thus axial load in external columns 1–2 and 22–23 is

$$N_{col(wind)} \times 21 = 21.25 \times 2.5 + 17.5 \times 6.0 + 17.5$$
$$\times 9.5 + 17.5 \times 13.0 + 17.5$$
$$\times 16.5 + 8.75 \times 20.0$$
$$= 53.125 + 105.0 + 166.25$$
$$+ 227.5 + 288.75 + 175.0$$
$$= 1015.625$$

Thus

$$N_{col(wind)} = 1015.625/21 = \pm 48.36 \text{ kN}$$

This value should be multiplied by 1.35 for the ULS, giving an additional vertical load of $48.36 \times 1.35 = \pm 65.29$ kN. This is of small order compared with the axial load from $1.35(G_k + Q_k)$.

8.2.7 Summary

The element sizes in Figure 8.2 appear to be adequate for design to EC2/NAD and can be used in the computer analysis for the detailed design. The estimated reinforcement percentages – 0.46 slab, 1.36 beam and 2.0 column – are reasonable and within the Code limits. As the vertical load moments dominate, the beam depth is maintained at a constant value of 600 mm for the six storeys.

8.3 PRELIMINARY DESIGN: CASE (ii) EC8/EC2

8.3.1 Loading–general

For the seismic design, the following adjustments have been made to the element sizes; see Figure 8.4:

beams h = 700 mm

b_w = 500 mm (L$_1$ to L$_3$)

h = 600 mm

b_w = 500 mm (L$_4$ to L$_6$)

columns   600 × 500 (G to L$_3$)

500 × 500 (L$_4$ to L$_6$)

The seismic coefficient for a linear analysis is dependent on the ratio of the design ground acceleration to the acceleration of gravity, the soil parameter, the ordinate of the design spectrum and the behaviour factor, thus reference should be made to EC8 and Chapter 7. For the purpose of this example, the seismic coefficient ε_0 is taken as 0.15.

The combination factor ψ for quasi-permanent loading is taken as 0.5 and the factor γ_1 (importance factor – used to express the importance of a building) is taken as unity.

The slab depth is maintained at 200 mm and thus G_k = 7.0 kN/m^2, as case (i). To allow for the self-weight of the beam ribs, columns and cladding, this value is increased by, say, 15%. Thus:

G_k = 7.0 × 1.15 = 8.05 kN/m^2

For a combination factor ψ of 0.5

ψQ = 5.0 × 0.5 = 2.5 kN/m^2

Thus total horizontal seismic action E per bay is:

E = (8.05 + 2.5) × 5 × 21 × 0.15 = 166.16 kN

Using an importance factor γ_1 = 1.0

± $\gamma_1 E$ = ± 166.16 kN per floor

The base shear is 166.16 × 6 = 996.98 kN (say, 1000 kN). This shear is distributed over the height of the building in a triangular form as shown in Figure 8.4. Using the 'portal' method, the column shears at ground to first floor level are:

external (1–2 and 22–23)

1000 × 1/6 = 166.66 kN

internal (8–9 and 15–16)

1000 × 1/3 = 333.33 kN

At first to second floor level, the column shears are:

external (2–3 and 23–24)

(1000 – 60.6) × 1/6 = 156.57 kN

internal (9–10 and 16–17)

(1000 – 60.6) × 1/3 = 313.13 kN

The corresponding column and beam moments are shown in Figure 8.5. The vertical load moments associated with the seismic shear moments are for $G + \psi Q$ = 10.55 kN/m^2. Thus the load per metre run of beam is 10.55 × 5 = 52.75 kN/m.

8.3.2 Preliminary beam design for flexure and shear

The beam moment at joint 9 approximates to:

M_{UDL} = 0.1 × 52.75 × 7^2 = 258.48 kN m

Total beam moment at joint 9 is

M_d = 258.48 + 690.66 = 949.14 kN

The maximum resistance moment,

M_u = 0.167 $f_{ck} b_w d^2$

For d = 700 – 60 = 640 mm and b_w = 500

M_u = 0.167 × 30 × 500 × 0.64^2

= 1026.05 kN m

From Figure 4.3,

$M_d/b_w d^2$ = 949.14 × 10^6/500 × 640^2 = 4.63

Thus

x/d = 0.42

and

x = 0.42 × 640 = 268.8 mm

Thus

d – 0.4 x = 640 – 0.4 × 268.8 = 532.48 mm

For f_{yk} = 400 N/mm^2,

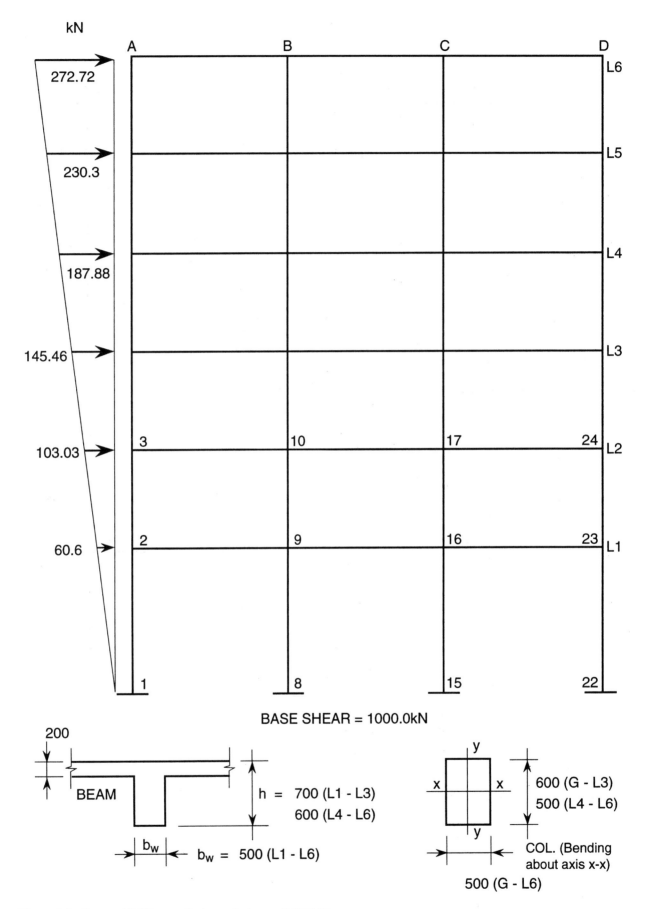

BASE SHEAR = 1000.0kN

Figure 8.4 Frame ABCD – preliminary design to EC8/EC2.

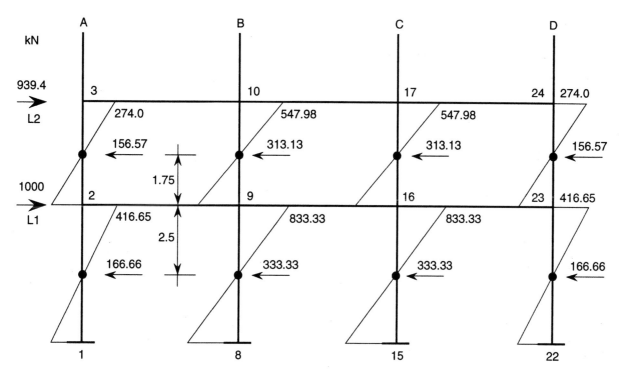

(a) Column Moments

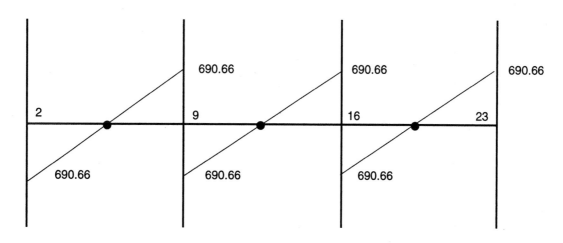

(b) Beam Moments

Beam Moments	=	166.66 x 2.5	=	416.66
		156.57 x 1.75	=	274.0
				690.66 kNm

Figure 8.5 Seismic analysis for beam and column moments.

$A_s = 949.14 \times 10^6/(400/1.15) \times 532.48$

$\quad = 5124.7 \text{ mm}^2$

This area of reinforcement could pose problems with detailing and reference should be made to Figure 4.15. Adopting a $10{-}25\varphi$ (4910 mm^2) and $2{-}20\varphi$ (628 mm^2), it is suggested that the $6{-}25\varphi$ bars are anchored beyond the joint and the $4{-}25$ and $2{-}16\varphi$ bars are anchored within the joint.

The steel percentage is $(4910 + 628)/500 \times 640 = 0.0173$. From Table 4.5, the maximum steel percentage for DC 'M' is given by:

$$p_{max} = 0.65(f_{cd}/f_{yd})\,(p^1/p) + 0.0015$$

With $p^1/p = 0.5$, say, then

$$p_{max} = 0.65 \times 17/347.8 \times 0.5 + 0.0015 = 0.174$$

This requirement is just satisfied.

The maximum beam shear force at joint 9 due to $(E + G + \psi Q)$ is

$V_d = 52.75 \times 3.5 + 258.48/7 + 2 \times 690.66/7$

$\quad = 184.63 + 36.93 + 197.33$

$\quad = 416.89 \text{ kN}$

In accordance with EC8, 40% of V_{Rd1}, as determined by EC2, represents the shear capacity of the concrete within the critical regions:

$0.4V_{Rd1}\,(V_{cd}) = 0.4\,[0.34\,(1.2 + 40 \times 0.0173)]$
$500 \times 640 \times 10^{-3}$

$\quad = 82.34 \text{ kN}$

Thus shear reinforcement is required to the value

$$V_{wd} = 416.89 - 82.34 = 334.55 \text{ kN}$$

This value will give a reasonable spacing of shear links, but it should be noted that Table 4.3 should be complied with. By inspection, V_{Rd2} is greater than V_d.

Preliminary design for bending and shear will not be taken further as the beam section is clearly adequate.

8.3.3 Preliminary column design

As with design for wind actions, the axial force in an external column for seismic actions is obtained by taking moments about the point of contraflexure for the portion of the frame above the load being considered. Thus:

$N_{col(seismic)} \times 21 = 272.72 \times 20 + 230.3 \times 16.5$

$\qquad + 187.88 \times 13.0 + 145.46$

$\qquad \times 9.5 + 103.03 \times 6.0$

$\qquad + 60.6 \times 2.5$

$\qquad = 5454.4 + 3799.95 + 2442.44$

$\qquad + 1383.2 + 619.8 + 151.5$

$\qquad = 13851.29$

Thus

$N_{col(seismic)} \quad = \pm 13851.29/21$

$\qquad\qquad = \pm 659.58 \text{ kN}$

The axial loads in the columns at ground to first floor level due to $(G + \psi Q)$ are:

external $(1{-}2$ and $22{-}23)$

$\quad 52.75 \times 7.0 \times 0.4 \times 6 = 886.2 \text{ kN}$

internal $(8{-}9$ and $15{-}16)$

$\quad 52.75 \times 7 \times 1.1 \times 6 = 2437.1 \text{ kN}$

The column moments arising from vertical load are of small order, and thus the following design moments and axial loads apply:

● *External*

$\quad N_d = 886.2 + 659.58 = 1545.78 \text{ kN}$

$\quad M_d = 416.65 \text{ kN m}$

and the parameters for use in the design charts for columns (see Appendix F) are

$\quad N_d/bh = 1545.78 \times 10^3/500 \times 600 = 5.15$

$\quad M_d/bh^2 = 420.31 \times 10^6/500 \times 600^2 = 2.34$

● *Internal*

$\quad N_d = 2437.1 \text{ kN}$

$\quad M_d = 833.33 \text{ kN m}$

and the corresponding parameters are

$\quad N_d/bh = 2437.1 \times 10^3/500 \times 600 = 8.12$

Table 8.2 Summary of beam and column moments, preliminary design

Loading condition	Beam moment, end 9 (kN m)	Beam shear, end 9 (kN)	Column axial load (kN)		Column moment (kN m)	
			External 1–2	*Internal 8–9*	*External 1–2*	*Internal 8–9*
$1.35G_k + 1.5Q_k$	440.68	377.72	1510.92	4154.34	42.94	negligible
$1.35(G_k + Q_k + W_k)$	509.54	380.43	1513.2	3981.7	97.36	87.24
$G_k + \psi Q_k + E$	949.14	416.89	1545.78	2437.1	833.33	416.65

$$M_d/bh^2 = 833.33 \times 10^6/500 \times 600^2$$

$$= 4.63$$

The values of the parameters are relatively low and thus the column size of 600×500 mm^2 at ground to first floor level is adequate.

8.4 PRELIMINARY DESIGN – SUMMARY

The preliminary designs for cases (i) and (ii) have concentrated on level 1 and the trial sizes appear to be adequate. If a member size is altered at a higher level, this should also be checked. It should be noted that the design for seismic actions assumes a vertical loading of 52.75 kN/m on the beams. Thus the EC2 loading of $1.35G_k + 1.5Q_k$ could involve higher beam moments and this should be checked at each floor level. A summary of the results for the preliminary analysis is given in Table 8.2.

8.5 DETAILED DESIGN

For a detailed design, the use of a computer package such as SAND is appropriate. The results for frame ABCD are not included here, but it was found that it was not necessary to adjust the member sizes adopted for the preliminary design. With the SAND analysis, two orthogonal frames were considered and the floor slabs were assumed to act as two-way spanning (aspect ratio = 7/5 = 1.4). The trapezoidal distribution of the permanent and variable loads will result in a reduced loading on the 7.0 m spans compared with the preliminary design for which one-way spanning was assumed. With the SAND analysis, it is also possible to include shear walls.

REFERENCES

SAND, *Structural Analysis and Design*, Fitzroy Computer Systems, 50 Fairmile Lane, Cobham, Surrey KT11 2DF.

Timoshenko S. and Woinowsky–Krieger S. (1959) *Theory of Plates and Shells*, McGraw-Hill, New York.

Appendix A

REINFORCEMENT AREAS

Tables of reinforcement areas for various situations are given in Tables A.1 to A.3. The values in the first two tables have been given to three significant figures according to the BSI recommendations.

Table A.1 Sectional areas of groups of bars $(mm^2)^a$

Bar size (mm)	Number of bars									
	1	2	3	4	5	6	7	8	9	10
6[b]	28.3	56.6	89.4	113	142	170	198	226	255	283
8	50.3	101	151	201	252	302	352	402	453	503
10	78.5	157	236	314	393	471	550	628	707	785
12	113	226	339	452	566	679	792	905	1020	1130
16	201	402	603	804	1010	1210	1410	1610	1810	2010
20	314	628	943	1260	1750	1890	2200	2510	2830	3140
25	491	982	1470	1960	2450	2950	3440	3930	4420	4910
32	804	1610	2410	3220	4020	4830	5630	6430	7240	8040
40	1260	2510	3770	5030	6280	7540	8800	10100	11300	12600
50[b]	1960	3930	5890	7850	9820	11800	13700	15700	17700	19600

[a]To three significant figures.
[b]Non-preferred sizes.

Table A.2 Sectional areas per metre width for various bar spacings $(mm^2/m)^a$

Bar size (mm)	Spacing of bars (mm)									
	75	100	125	150	175	200	225	250	275	300
6[b]	377	283	226	189	162	142	126	113	103	94.3
8	671	503	402	335	287	252	224	201	183	168
10	1050	785	628	523	449	393	349	314	285	262
12	1510	1130	905	754	646	566	503	452	411	377
16	2680	2010	1610	1340	1150	1010	894	804	731	670
20	4190	3140	2510	2090	1800	1570	1400	1260	1140	1050
25	6550	4910	3930	3270	2810	2450	2180	1960	1790	1640
32	10700	8040	6430	5360	4600	4020	3570	3220	2920	2680
40	16800	12600	10100	8380	7180	6280	5580	5030	4570	4190
50[b]	26200	19600	15700	13100	11200	9820	8730	7850	7140	6540

[a]To three significant figures.
[b]Non-preferred sizes.

Table A.3 Fabric to BS 4483[a]. British Standard preferred meshes in stock size sheets 4.8 m long by 2.4 m wide

British Standard reference	Longitudinal wires			Cross wires			Mass	
	Size (mm)	Pitch (mm)	Area (mm²/m)	Size (mm)	Pitch (mm)	Area (mm²/m)	(kg/m²)	(kg/sheet)
Square mesh fabric								
A 393	10	200	393	10	200	393	6.16	70.96
A 252	8	200	252	8	200	252	3.95	45.50
A 193	7	200	193	7	200	193	3.02	34.79
A 142	6	200	142	6	200	142	2.22	25.57
A 98	5	200	98	5	200	98	1.54	17.74
Structural fabric								
B 1131	12	100	1131	8	200	252	10.9	125.57
B 785	10	100	785	8	200	252	8.14	93.77
B 503	8	100	503	8	200	252	5.93	68.31
B 385	7	100	385	7	200	193	4.53	52.19
B 283	6	100	283	7	200	193	3.73	42.97
B 196	5	100	196	7	200	193	3.05	35.14
Long mesh fabric								
C 785	10	100	785	6	400	70.8	6.72	77.41
C 636	9	100	636	6	400	70.8	5.55	63.94
C 503	8	100	503	5	400	49	4.34	50.00
C 385	7	100	385	5	400	49	3.41	39.28
C 283	6	100	283	5	400	49	2.61	30.07

[a]Fabric is produced from cold-drawn wire with a characteristic strength of not less than 460 N/mm².

Appendix B

GUIDELINES FOR PRELIMINARY DESIGN

B.1 INTRODUCTION

At the conceptual (preliminary) design stage, it is important that the structural options available can be quantified with reasonable accuracy in order to assess their implications with regard to functional requirements and cost. For conventional building frameworks, this will generally involve the proportioning of slabs, beams and columns. These data can, of course, be utilized as input for computer analysis at the detailed design stage. The following procedure is suggested for low-rise building frameworks (say, up to six storeys) under the general headings 'loading', 'member analysis' and 'section analysis'. For convenience, Figures 3.3 and 4.3 and Tables 1.6 and 4.13 are included in these guidelines (as Figures B.1 and B.2 and Tables B.2 and B.1).

B.2 LOADING

Having established the overall dimensions of a structure from functional requirements, that is, bay sizes, beam spans, storey heights and so on, the permanent loads of the slabs and beams can be estimated from the span to effective depth ratios (L/d) given in Table B.1. For slabs, use the L/d ratios for lightly stressed members, and for beams, the L/d ratios for highly stressed members. Note that for spans exceeding 7.0 m, the modification factor $7/L_{eff}$ is required. Table B.1 is based on the assumption that $f_{yk} = 400$ N/mm² and thus, if f_{yk} is 460 N/mm², a modification factor of 0.87 is suggested. The relevant NAD should be consulted for the appropriate cover to meet durability requirements, minimum dimensions for fire resistance and variable (imposed) loadings. In these guidelines it will be assumed that, in general, Q_k will not exceed 5 kN/m².

Table B.1 Basic ratios of span/effective depth for reinforced concrete members without axial compression (table 4.14 of EC2)[a]

Structural system	Concrete highly stressed	Concrete lightly stressed
1. Simply supported beam, one- or two-way spanning simply supported slab	18	25
2. End span of continuous beam or one-way continuous slab or two-way spanning slab continuous over one long side	23	32
3. Interior span of beam or one-way or two-way spanning slab	25	35
4. Slab supported on columns without beams (flat slab) – based on longer span	21	30
5. Cantilever	7	10

[a]For slabs, use L_{eff}/d ratios for lightly stressed members, and for beams, use L_{eff}/d ratios for highly stressed members.

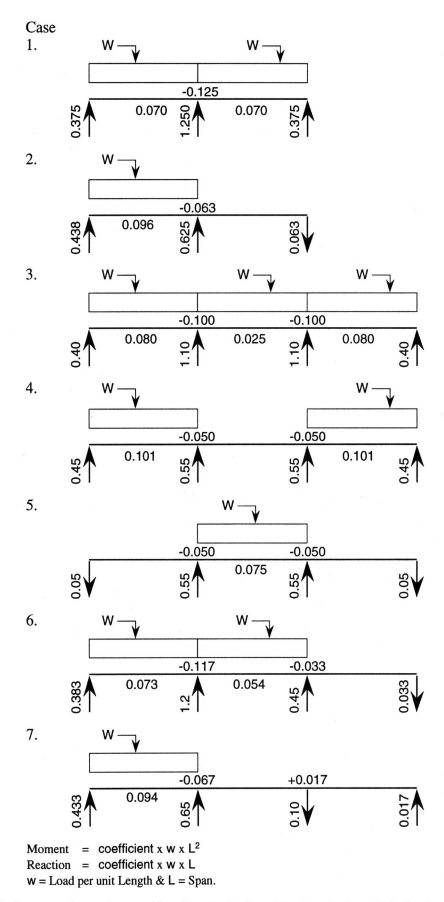

Moment   =   coefficient x w x L²
Reaction =   coefficient x w x L
w = Load per unit Length & L = Span.

Figure B.1   Equal-span continuous beams with uniformly distributed loads – elastic analysis (load cases 1–13).

Case

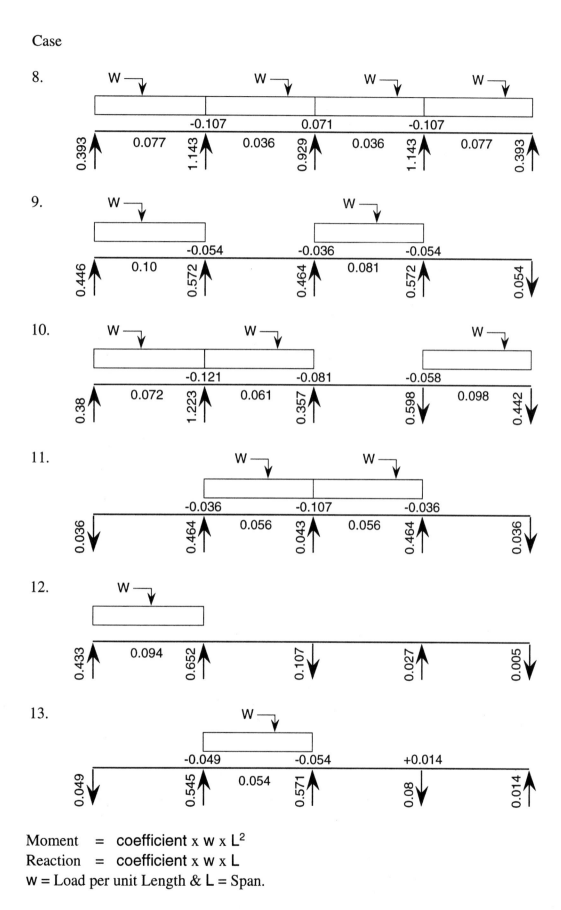

Moment = coefficient x w x L²
Reaction = coefficient x w x L
w = Load per unit Length & L = Span.

Figure B.1 *Continued*

The above procedure is not appropriate for the seismic design of frameworks, and in this case the L/d ratio for beams should be in the range 10 to 12.

B.3 MEMBER ANALYSIS

This should be carried out for the ultimate limit state, and the design bending moment M_{sd} and shear force V_{sd} for beams and slabs may be obtained from Figure B.1. These are elastic coefficients (no redistribution) and the possibility of adjustment of moments should be considered at the detailed design stage. Alternatively, a plastic analysis may be adopted for slabs; see Appendices D and E. The vertical loads on columns are obtained from the floor areas they support. If there are no shear walls and the columns are to be designed to resist wind load moments, then this can be allowed for (including moments arising from vertical load) by increasing the vertical (axial) load on the column by 50% (25% if moments arise from vertical load only). Alternatively, moments in columns can be assessed rapidly from the procedure given in section 3.4.3. For seismic design, the base shear is obtained from the equivalent static analysis, and then distributed in triangular form over the height of the building. Again, an approximation to the beam and column moments may be obtained from section 3.4.3.

B.4 SECTION ANALYSIS

B.4.1 General

Select a suitable steel and concrete grade. The concrete grade should be related to both **strength** and **durability** requirements and, for normal design purposes, C30 or C35 will generally be adequate.

The properties of concrete are brought together in Table B.2. For steel, $f_{yk} = 460$ N/mm² is the norm in the UK, but there are variations across Europe, and this should be noted.

B.4.2 Slabs and beams

(a) Flexure

The neutral axis factor x is obtained from Figure B.2 and thus the steel area can be obtained from

$$A_s = M_{sd}/(f_{yk}/\gamma_s) \, (d - 0.4 \, x)$$

and

$$p_1 = A_s/b_w d$$

Note the limitations in reinforcement ratio ($p = A_s/b_w d$) given in Tables 4.1 and 4.8. For design to EC2 (not EC8) the maximum flexural capacity at ULS for single reinforced sections in flexure is

$$M_u = 0.167 \, f_{ck} \, b_w d^2$$
$$(f_{ck} \leqslant 35 \text{ N/mm}^2, \, x/d = 0.45)$$

$$M_u = 0.128 \, f_{ck} \, b_w d^2$$
$$(f_{ck} > 35 \text{ N/mm}^2, \, x/d = 0.35)$$

For seismic design, M_u values based on limiting values of p are given in Table 4.8.

(b) Shear

The basic shear stress τ_{Rd} is obtained from Table B.2, noting that in the NAD the maximum value of f_{ck} that can be used in the determination of τ_{Rd} should be taken as 40 N/mm². Note that, in Chapter 4, V_{Rd1} is given as $\tau_{Rd} \, k \, (1.2 + 40 \, p_1) \, b_w d$.

Table B.2 Summary of properties of concrete (five grades only) all related to the characteristic compressive cylinder strength of concrete (f_{ck}, N/mm²) at 28 days

Strength class of concrete	C20/25	C25/30	C30/37	C35/45	C40/50
f_{ck}	20	25	30	35	40
f_{ctm}	2.2	2.6	2.9	3.2	3.5
$f_{ctk,0.05}$	1.5	1.8	2.0	2.2	2.5
$f_{ctk,0.95}$	2.9	3.3	3.8	4.2	4.6
τ_{Rd}	0.26	0.30	0.34	0.37	0.41
E_{cm}	29	30.5	32	33.5	35
f_{bd} plain bars	1.1	1.2	1.3	1.4	1.5
f_{bd} high-bond bars $\varphi \geqslant 34$ mm	2.3	2.7	3.0	3.4	3.7

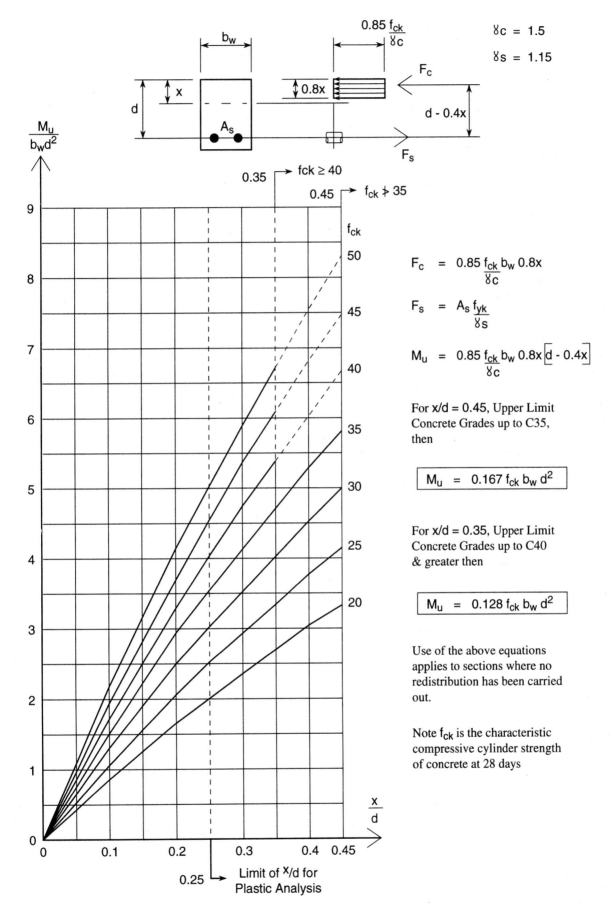

$$F_c = \frac{0.85\, f_{ck}\, b_w\, 0.8x}{\gamma_c}$$

$$F_s = \frac{A_s\, f_{yk}}{\gamma_s}$$

$$M_u = \frac{0.85\, f_{ck}\, b_w\, 0.8x}{\gamma_c}\left[d - 0.4x\right]$$

For x/d = 0.45, Upper Limit Concrete Grades up to C35, then

$$\boxed{M_u = 0.167\, f_{ck}\, b_w\, d^2}$$

For x/d = 0.35, Upper Limit Concrete Grades up to C40 & greater then

$$\boxed{M_u = 0.128\, f_{ck}\, b_w\, d^2}$$

Use of the above equations applies to sections where no redistribution has been carried out.

Note f_{ck} is the characteristic compressive cylinder strength of concrete at 28 days

Figure B.2 EC2 design chart for flexure.

Table B.3 If $V_{sd} > V_{Rd1}$ then one should have $V_{sd} < V_{Rd2}$ given below

f_{ck}	V_{Rd2}
20	3.6 $b_w d$
25	4.31 $b_w d$
30	4.95 $b_w d$
35	5.51 $b_w d$
40	6.0 $b_w d$

If V_{sd} exceeds V_{Rd1} then check that V_{sd} is less than V_{Rd2} given in Table B.3. For $V_{sd} > V_{Rd1} < V_{Rd2}$ provide shear reinforcement links, given by

$$V_{sd} - V_{Rd1} = V_{wd} = (A_{sw}/s) \times 0.9 \, d \, f_{ywd}$$

Considerations of torsion can normally be omitted at the preliminary design stage.

B.4.3 Columns

To avoid the necessity of considering slenderness effects, limit the ratio of storey height to least lateral dimension of the columns to 12. Allow for bending effects as in section B.3 by increasing the axial load by 25–50%. Working in terms of axial load only, the design ultimate load capacity of a section is:

$$N_{ud} = \alpha f_{cd} A_c + f_{yd} A_s$$

With $\alpha = 0.85$, $f_{cd} = f_{ck}/1.5$ and $f_{yd} = f_{yk}/1.15$

$$N_{ud} = 0.57 \, f_{ck} A_c + 0.87 \, f_{yk} A_s$$

With $p = A_s/A_c$

$$N_{ud} = A_c \, (0.57 \, f_{ck} + 0.87 \, p \, f_{yk})$$

Table B.4 Ultimate axial load capacity of stocky columns N_{ud} (kN): (1) $0.57 f_{ck} A_c$; (2) $0.87 p f_{yk} A_c$; (3) = (1) + (2), and $N_{bal} = 0.4 f_{cd} A_c$ (kN) and $f_{ck} = 30$ N/mm^2 and $f_{yk} = 460$ N/mm^2

Column size (mm × mm)		0.01	0.02	0.03	0.04	N_{bal} (kN) = 0.4 f_{cd} A_c
250 × 250	(1)	1069	1069	1069	1069	
	(2)	250	500	750	1000	500
	(3)	1319	1569	1819	2069	
300 × 300	(1)	1539	1539	1539	1539	
	(2)	360	720	1080	1440	720
	(3)	1899	2259	2619	2979	
350 × 350	(1)	2095	2095	2095	2095	
	(2)	490	980	1470	1960	980
	(3)	2585	3075	3565	4055	
400 × 400	(1)	2736	2736	2736	2736	
	(2)	640	1280	1920	2560	1280
	(3)	3376	4016	4656	5296	
450 × 450	(1)	3463	3463	3463	3463	
	(2)	810	1620	2430	3240	1620
	(3)	4273	5083	5893	6703	
500 × 500	(1)	4275	4275	4275	4275	
	(2)	1000	2000	3000	4000	2000
	(3)	5275	6275	7275	8275	
550 × 550	(1)	5173	5173	5173	5173	
	(2)	1210	2420	3630	4840	2420
	(3)	6383	7593	8803	10013	
600 × 600	(1)	6156	6156	6156	6156	
	(2)	1440	2880	4320	5760	2880
	(3)	7596	9036	10476	11916	

The column group header "p = A_s/A_c" spans the 0.01, 0.02, 0.03, 0.04 columns.

Table B.5 Values of limiting axial force N_d (kN) for the three levels of ductility with $f_{ck} = 30$ N/mm². For grades 20, 25 and 35, multiply by 0.67, 0.83 and 1.17 respectively

Column (mm × mm)	N_d (kN)		
	Low (L)	Medium (M)	High (H)
300 × 300	1350	1170	990
350 × 350	1838	1593	1348
400 × 400	2400	2080	1760
450 × 450	3038	2633	2228
500 × 500	3750	3250	2750
550 × 550	4538	3933	3328
600 × 600	5400	4680	3960
650 × 650	6388	5493	4648
700 × 700	7350	6370	5390
750 × 750	8438	7313	6188
800 × 800	9600	8320	7040
850 × 850	10838	9393	7948
900 × 900	12150	10530	8910
950 × 950	13538	11733	9928
1000 × 1000	15000	13000	11000

Using the above equation, Table B.4 has been drawn up for $p = 0.01$ to 0.04 with $f_{ck} = 30$ N/mm². For preliminary design, working to the upper limit $p = 0.04$ is not advisable and $p = 0.02$ is suggested. To maximize the ultimate moment capacity of a section, the maximum axial load that should be applied (N_{bal}) is $0.4f_{cd}A_c$ (see EC2 cl. 4.3.5.6.3). These values are also tabulated.

For seismic design, EC8 puts limitations on acting axial force N_d as follows:

low ductility (L) $N_d = 0.75\, f_{cd}A_c$

medium ductility (M) $N_d = 0.65\, f_{cd}A_c$

high ductility (H) $N_d = 0.55\, f_{cd}A_c$

Thus for $f_{ck} = 30$ N/mm² and $\gamma_c = 1.5$, the values of N_d are $15A_c$, $13A_c$ and $11A_c$ respectively. Tabulated values of N_d (kN) are given in Table B.5.

Appendix C

RATIOS OF DESIGN BENDING MOMENTS (EC2/BS 8110)

Using the bending moment coefficients in Figure 3.3, the ratios of design bending moment coefficients (EC2/BS 8110) are evaluated below for a two-span and four-span beam. It is assumed that the characteristic permanent load G_k is equal to the characteristic variable load Q_k. The partial safety factors for ultimate limit state are $\gamma_{Gk} = 1.35$ and $\gamma_{Qk} = 1.5$ for EC2 and $\gamma_{Gk} = 1.4$ and $\gamma_{Qk} = 1.6$ for BS 8110.

C.1 TWO-SPAN BEAM – EC2 LOAD CASES (see Figure C.1 (a))

Case (i)

We have

$$M_B = 0.125 \ (1.35 \ G_k + 1.5 \ Q_k) \ L^2$$

For $G_k = Q_k$

$$M_B = 0.125 \times 2.85 G_k \times L^2 = 0.356 \ G_k L^2$$

$$M_{AB} = 0.07 \ (1.35 \ G_k + 1.5 \ Q_k) \ L^2$$

For $G_k = Q_k$

$$M_{AB} = 0.07 \times 2.85 G_k \times L^2 = 0.1995 \ G_k L^2$$

Case (ii)

We have

$$M_{AB} = 0.07 \times 1.35 G_k \times L^2 + 0.096 \times 1.5 Q_k \times L^2$$
$$= 0.0945 \ G_k L^2 + 0.144 \ Q_k L^2$$

For $G_k = Q_k$

$$M_{AB} = (0.0945 + 0.144) \ G_k L^2 = 0.2385 \ G_k L^2$$

C.2 TWO-SPAN BEAM – BS 8110 LOAD CASES (see Figure C.1(b))

Case (i)

We have

$$M_B = 0.125 \ (1.4 \ G_k + 1.6 \ Q_k) \ L^2$$

For $G_k = Q_k$

$$M_B = 0.125 \times 3.0 G_k \times L^2 = 0.375 \ G_k L^2$$

Case (ii)

We have

$$M_{AB} = 0.07 \times 1.0 \ G_k \times L^2$$
$$+ 0.096(0.4 \ G_k + 1.6 \ Q_k)L^2$$

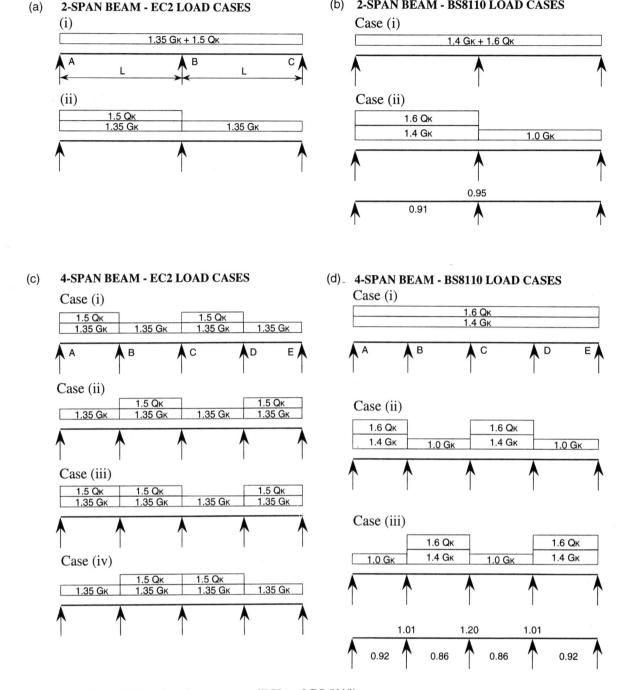

Figure C.1 Ratios of design bending moments (EC3 and BS 8110).

For $G_k = Q_k$

$$M_{AB} = (0.07 + 0.096 \times 2.0) \; G_k L^2$$
$$= 0.262 \; G_k L^2$$

Thus the ratios of design bending moments (EC2/BS 8110) for equal spans and $G_k = Q_k$ are:

$$M_B = 0.356/0.375 = 0.95$$

$$M_{AB} = 0.2385/0.262 = 0.91$$

C.3 FOUR-SPAN BEAM – EC2 LOAD CASES (see Figure C.1(c))

For maximum moment in span AB, use case (i)

$$M_{AB} = 0.077 \times 1.35 \; G_k \times L^2 + 0.1 \times 1.5 \; Q_k \times L^2$$
$$= 0.104 \; G_k L^2 + 0.15 \; Q_k L^2$$

For $G_k = Q_k$

$$M_{AB} = (0.104 + 0.15) \; G_k L^2 = 0.254 \; G_k L^2$$

For maximum moment at support B, use load case (iii)

$$M_B = 0.107 \times 1.35 \; G_k \times L^2 + 0.121 \times 1.5 \; Q_k \times L^2$$

$$= 0.144 \; G_k L^2 + 0.182 \; Q_k L^2$$

For $G_k = Q_k$

$$M_B = (0.144 + 0.182) \; G_k L^2 = 0.326 \; G_k L^2$$

For maximum moment in span BC, use load case (ii)

$$M_{BC} = 0.036 \times 1.35 \; G_k \times L^2 + 0.081 \times 1.5 \; Q_k \times L^2$$

$$= 0.0486 \; G_k L^2 + 0.1215 \; Q_k L^2$$

For $G_k = Q_k$

$$M_{BC} = (0.0486 + 0.1215) \; G_k L^2 = 0.1701 \; G_k L^2$$

For maximum moment at support C, use load case (iv)

$$M_C = 0.071 \times 1.35 \; G_k \times L^2 + 0.107 \times 1.5 \; Q_k \times L^2$$
$$= 0.09585 \; G_k L^2 + 0.1605 \; Q_k L^2$$

For $G_k = Q_k$

$$M_c = (0.09585 + 0.1605) \; G_k L^2 = 0.256 \; G_k L^2$$

C.4 FOUR-SPAN BEAM – BS 8110 LOAD CASES (see Figure C.1(d))

For maximum moment in span AB, use load case (ii)

$$M_{AB} = 0.077 \times 1.0 \; G_k \times L^2$$
$$+ 0.10 \; (0.4 \; G_k + 1.6 \; Q_k)L^2$$

For $G_k = Q_k$

$$M_{AB} = 0.077 \times G_k \times L^2 + 0.1 \times 2.0 \times G_k \times L^2$$
$$= 0.277 \; G_k L^2$$

For maximum moment at support B, use load case (i)

$$M_B = 0.107 \; (1.4 \; G_k + 1.6 \; Q_k) \; L^2$$

For $G_k = Q_k$

$$M_B = 0.107 \times 3.0 \; G_k \times L^2 = 0.321 \; G_k L^2$$

For maximum moment in span BC, use load case (iii)

$$M_{BC} = 0.036 \times 1.0 \; G_k \times L^2 + 0.081 \; (1.6 \; Q_k + 0.4 \; G_k) \times L^2$$

For $G_k = Q_k$

$$M_{BC} = 0.036 \; G_k L^2 + 0.081 \times 2.0 \; G_k \times L^2 = 0.198 \; G_k L^2$$

For maximum moment at support C, use load case (i)

$$M_c = 0.071 \; (1.4 \; G_k + 1.6 \; Q_k) \times L^2$$

For $G_k = Q_k$

$$M_c = 0.071 \times 3.0 \; G_k \times L^2 = 0.213 \; G_k L^2$$

The ratios of design bending moments (EC2/BS 8110) for equal spans and $G_k = Q_k$ are:

$$M_{AB} = 0.254/0.277 = 0.92$$

$$M_B = 0.326/0.321 = 1.02$$

$$M_{BC} = 0.1701/0.198 = 0.86$$

$$M_C = 0.256/0.213 = 1.20$$

Appendix D

YIELD-LINE ANALYSIS – DESIGN FORMULAE

D.1 INTRODUCTION

The object of a yield-line analysis is to postulate a yield-line pattern from which the ultimate moment can be determined by:

- considering the equilibrium of the slab elements, or
- using the work equation (Wood, 1967).

The inherent difficulty of a yield-line analysis is to determine the yield-line pattern that will give a minimum value of the ultimate load for a given arrangement of reinforcement. In general, an upper bound solution is obtained (correct or unsafe) (Wood, 1967). However, owing to membrane effects, it can be shown that much higher ultimate loads than those predicted by a yield-line analysis can be obtained. Thus the yield-line theory may be used in the design office if a reasonable attempt is made to find the minimum ultimate load. The general procedure for the analysis of a square slab with simply supported edges is outlined below (case 1, Figure D.1), and the formulae only will be given for all other cases (cases 2–6, Figure D.2). A comprehensive tabulation of design formulae for simple slabs of various shapes, including flat slab floors, is given in the Unesco (1971) manual.

D.2 SQUARE SLAB, SIMPLY SUPPORTED, UNIFORMLY DISTRIBUTED LOADING

The assumed yield-line pattern is shown in Figure D.1 and further assumptions are as follows:

1. The external work done by the slab elements under unit displacement ($\delta = 1$ at point O) is equated to the components of internal work expressed as (moment per unit length along axis of rotation) × (length of projection) × (rotation about the axis under consideration).
2. Each yield line is considered as an infinite succession of straight lines disposed stepwise (Unesco, 1971).
3. The ratio of the ultimate moments per unit length in the x and y directions (orthogonal axes) is given by $\mu = m_y/m_x$.
4. The area of tensile reinforcement at any point or in any direction is such that the value of x/d does not exceed 0.25; see Chapter 3.

For unit displacement of the slab at O, it can then be considered as four plate elements, the centre of gravity of each plate element undergoing a displacement of one-third. Then

external work for plate OAB

$$= w \times (L^2/4) \times (1/3)$$
$$= wL^2/12$$

and thus

total external work $= 4 \times wL^2/12 = wL^2/3$

For plate OAB, there is rotation about the y axis only. The rotation θ of the slab is $2/L$, the moment per unit length is m_x and the length of the projection about the y axis (AB) is L. Thus

internal work for plate OAB

$$= m_x \times L \times (2/L)$$
$$= 2\,m_x$$

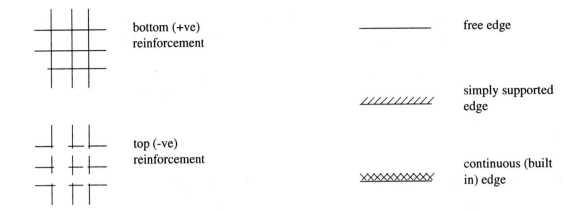

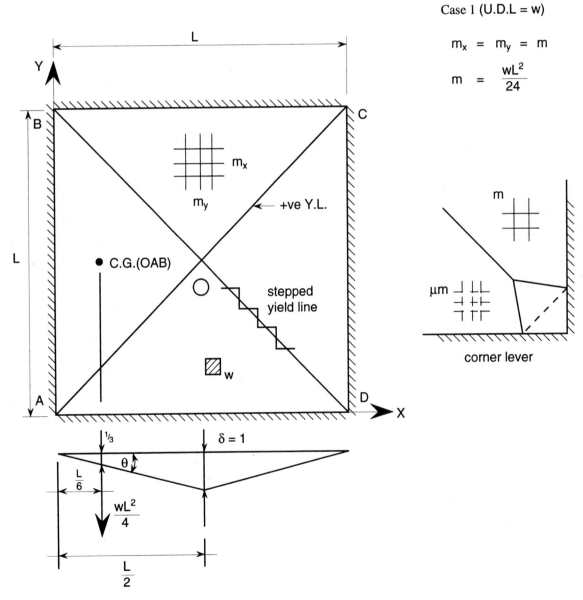

Case 1 (U.D.L = w)

$$m_x = m_y = m$$

$$m = \frac{wL^2}{24}$$

corner lever

Figure D.1 Yield-line analysis – general notation and case 1.

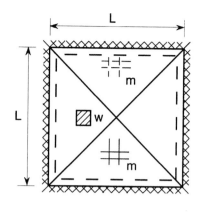

Case 2   Square Slab, continuous edges, Isotropic reinforcement - Top, Bottom, U.D.L. = W

$$m = \frac{wL^2}{48}$$

——— +ve Yield Line

– – – -ve Yield Line

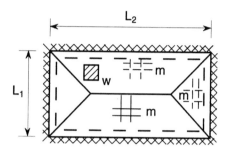

Case 3   Rectangular Slab, continuous edges, Isotropic reinforcement μm (top) and m (bottom), U.D.L. - W

$$m = \frac{W \cdot L_1 \cdot L_2}{8(1+\mu)\left[1 + \dfrac{L_2}{L_1} + \dfrac{L_1}{L_2}\right]}$$

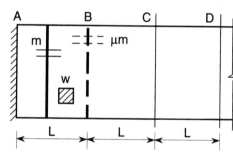

Case 4   1-way continuous slab m (+ve) and μm (-ve), U.D.L. = W, end span AB

$$m = \frac{wL^2}{2}\left[\frac{(1+\mu)^{1/2} - 1}{\mu}\right]^2$$

μ	m	μm
0.5	$\dfrac{wL^2}{10}$	$\dfrac{wL^2}{20}$
1.0	$\dfrac{wL^2}{11.6}$	$\dfrac{wL^2}{11.6}$
1.5	$\dfrac{wL^2}{13.4}$	$\dfrac{wL^2}{9}$
2.0	$\dfrac{wL^2}{15}$	$\dfrac{wL^2}{7.5}$

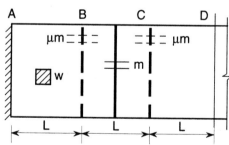

Case 5   1-way continuous slab m (+ve) and μm (-ve), U.D.L. = W, internal span BC

$$m = \frac{wL^2}{8(1+\mu)}$$

μ	m	μm
1	$\dfrac{wL^2}{16}$	$\dfrac{wL^2}{16}$
1.5	$\dfrac{wL^2}{20}$	$\dfrac{wL^2}{13.3}$
2.0	$\dfrac{wL^2}{24}$	$\dfrac{wL^2}{12}$

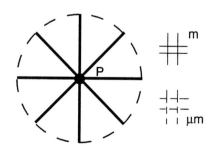

Case 6   Fan pattern, Isotropic reinforcement μm (top) and m (bottom), concentrated load P

$$P = 2\pi m (1 + \mu)$$

$$= 4\pi m \text{ for } \mu = 1$$

For application of fan pattern to flat slabs, see references

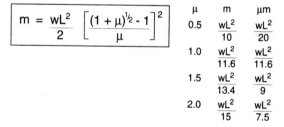

Figure D.2   Yield-line analysis – design formulae for cases 2–6.

The internal work for plate OCD is $2m_x$ and that for plates OBC and OAD is $2m_y$. Thus

total internal work $= 2 (2 m_x + 2 m_y)$

Thus

$m_x + m_y = wL^2/12$

If $m_x = m_y = m$, then

$m = wL^2/24$

or

$w = 24 m/L^2$

It must be emphasized that the yield-line pattern shown in Figure D.1 is merely a postulation and it is necessary to examine alternative patterns which may result in a lower value of w than $24m/L^2$. It is possible for a yield line to fork before it reaches the slab edge (Wood, 1967; Jones and Wood, 1967), forming what is known as a corner lever (see Figure D.1). Yield-line patterns that have corner levers are more critical, but the error involved is small for square slabs and is related to the ratio of the top to bottom reinforcement $\mu m/m = \mu$. The following results apply to a square slab with uniformly distributed loading (w):

μ	m
0	$wL^2/22$
0.25	$wL^2/23$
0.50	$wL^2/23.6$
1.0	$wL^2/24$

In many cases, the general form of the work equation can be expressed as:

$m = u/v$

where u and v are functions of x. Thus

$dm/dx = (v \, du/dx - u \, dv/dx)/v^2$

Putting

$dm/dx = 0$

then

$u/v = (du/dx)/(dv/dx)$

This expression has a number of applications, e.g. to rectangular slabs (see Wood, 1967; Jones and Wood, 1967).

REFERENCES

Jones L.L. and Wood R.H. (1967) *Yield Line Analysis of Slabs*, Chatto and Windus/Thames and Hudson, London.

Unesco (1971) *Reinforced Concrete, An International Manual*, English translation, Butterworths, London.

Wood R.H. (1967) *Plastic and Elastic Design of Slabs and Plates*, Thames and Hudson, London.

Appendix E

THE STRIP METHOD
WITH NUMERICAL APPLICATION

E.1 INTRODUCTION

In section 3.4.1(a), it was indicated that the strip method (after Hillerborg, 1975) is a useful approach to the analysis of slabs of certain configurations at the ultimate limit state. The strip method is a design approach in which the load is carried by a series of strips assumed to act like beams. The strips are reinforced along their full length like beams carrying the same load. Thus it is possible to design a slab with variable reinforcement, in contrast to the yield-line theory, which, in general, is used to analyse a slab with an assumed arrangement of uniform reinforcement. The slab may be divided into a series of x and y strips, thus reducing the equilibrium equation of a slab element to the form

$$\partial^2 M_x/\partial x^2 + \partial M_y/\partial_y{}^2 = -(W_x + W_y) = -W$$

It is assumed that plasticity allows the deflection of perpendicular strips to become compatible. Division of the load into x and y strips is achieved by introducing load dispersion lines (Figure E.1(a)). This implies that trapezoidally loaded areas are formed, which means that continuously varying reinforcement is required, and this is not practicable. Wood and Armer (1970) suggest the adoption of discontinuity lines as shown in Figure E.1(b) and thus the reinforcement may be placed in bands of convenient width. Continuity over supports is dealt with by introducing zero moment lines; see Figure E.1(c). The position of the zero moment lines should be chosen such that the ratio of support to span moment does not depart too far from that obtained by an elastic analysis. A large departure may result in serviceability problems; see Chapter 4. The general approach is considered in the following outline design for the ultimate limit state.

E.2 EXAMPLE OF USE OF STRIP METHOD

A reinforced concrete tank wall (see Figure E.2(a)) built in on three sides and free on the fourth is subjected to an equivalent fluid pressure of 5 kN/m² per metre height of wall. γ_f will be taken as 1.5 for ultimate limit state.

E.2.1 Loading (ULS)

The first step is to choose the position of the load dispersion and zero moment lines. This choice should be made bearing in mind the elastic distribution of bending moments at service loads, otherwise crack widths at service loads may be excessive. Timoshenko and Woinowsky-Krieger (1959) give tabulated data for elastic analysis. Two typical strips will be considered; see Figure E.2(a). Strip 1–1 may be taken as a beam built in at both ends, and an elastic analysis would give the point of contraflexure at a distance of $0.21 \times 6 = 1.26$ m from the supports. Thus for the ultimate limit state, the distance of the zero moment line from the support should not be too far from 1.26 m and a value of 1.0 m has been assumed. The centre line of strip 1–1 (1.0 m wide) has a loading intensity of $5 \times 5 = 25$ kN/m², and this will be taken as constant over the full width of the strip; see Figure E.2(b). The loading on strip 2–2 is shown in Figure E.2(c). In order to maintain equilibrium, a reaction R_2 is required at the free edge (for which additional reinforcement is required).

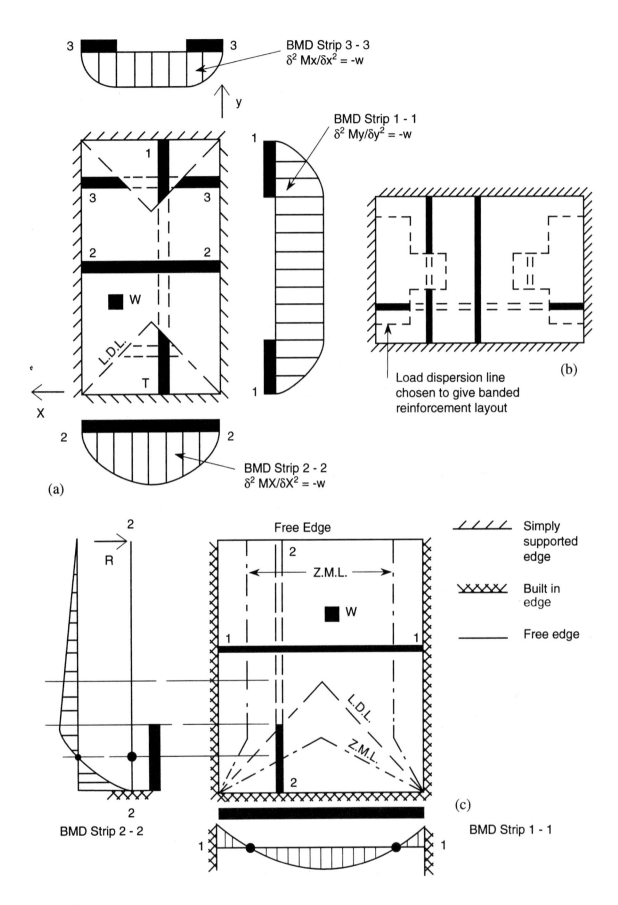

Figure E.1 The basis of the strip method.

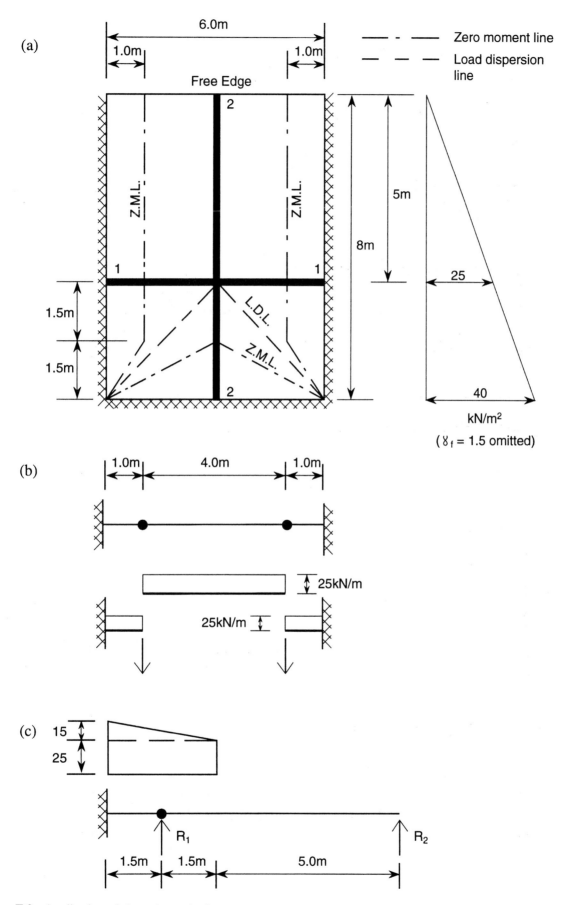

Figure E.2 Application of the strip method.

E.2.2 Member analysis (ULS)

Strip 1–1

Between the zero moment lines, the strip is designed as a simple beam. Thus the maximum bending moment M and reaction at the zero moment line are

$$M = 0.125 \times 25 \times 1.5 \times 4^2 = 75 \text{ kN m/m}$$

$$V = 0.5 \times 25 \times 1.5 \times 4 = 75 \text{ kN}$$

The moment M and shear V at the slab edge are

$$M = 75 \times 1.0 + 25 \times 1.0 \times 0.5 = 87.5 \text{ kN m/m}$$

$$V = 75 + 1.0 \times 25 \times 1.5 = 112.5 \text{ kN}$$

Strip 2–2

$$R_2 = 25 \times 1.5 \times 1.5 \times 0.75/6.5 + 3.75 \times 1.5$$
$$\times 1.5 \times 0.5/6.5$$
$$= 7.14 \text{ kN}$$

$$R_1 = 25 \times 1.5 \times 1.5 \times 5.75/6.5 + 3.75 \times 1.5$$
$$\times 1.5 \times 6.0/6.5$$
$$= 57.55 \text{ kN}$$

The bending moment at various positions along strip 2–2 can now be evaluated. As with strip 1–1, the strip is assumed to be uniformly loaded across its metre width. Thus

$$M_5 = R_2 \times 5 = 7.14 \times 5 = 35.7 \text{ kN m/m}$$

$$M_8 = 57.55 \times 1.5 + 32.5 \times 1.5 \times 1.5 \times 0.75$$
$$+ 3.75 \times 1.5 \times 1.5 \times 0.5$$
$$= 86.33 + 54.84 + 4.22$$
$$= 145.39 \text{ kN m/m}$$

The shear force at the base of the wall for strip 2–2 is

$$V_8 = 57.55 + 32.5 \times 1.5 \times 1.5 + 3.75 \times 1.5 \times 1.5$$
$$= 139.11 \text{ kN}$$

E.2.3 Section analysis (ULS)

For brevity, M_8 and V_8 only will be considered with the following data: $h = 250$ mm, $d = 210$ mm, $f_{ck} = 35$ N/mm^2 and $f_{yk} = 400$ N/mm^2. From Figure 4.3 (or B.2)

$$M_8/b_w d^2 = 145.39 \times 10^6/10^3 \times 210^2 = 3.3$$

Thus for $f_{ck} = 35$ N/mm^2, $x/d = 0.23$. (The strip method is a plastic analysis and thus the upper limit of x/d should be taken as 0.25.) Thus

$$x = 0.23 \times 210 = 48.3$$

$$d - 0.4\, x = 190.68 \text{ mm}$$

Thus

$$A_s = M_8/(f_{yk}/\gamma_s)(d - 0.4\, x)$$
$$= 145.39 \times 10^6 \times 1.15/400 \times 190.68$$
$$= 2192 \text{ mm}^2/\text{m} \quad (20\, \varphi - 140 = 2243 \text{ mm}^2/\text{m})$$

and

$$p = 2243/210 \times 10^3 = 0.0107$$

From Table 1.6 (or B.2), the basic shear stress τ_{Rd} is 0.37 N/mm^2 for $f_{ck} = 35$ N/mm^2. V_{Rd1} is given by equation (4.2) as

$$V_{Rd1} = \tau_{Rd}\, k\, (1.2 + 40\, p_1)\, b_w d$$

and

$$k = 1.6 - 0.21 = 1.39$$

$$p_1 = 0.0107$$

Thus

$$V_{Rd1} = 0.37 \times 1.39\, (1.2 + 40 \times 0.0107) \times 210$$
$$= 175.83 \text{ kN}$$

Thus the section is adequate in shear.

A serviceability check should be carried out in accordance with Chapter 4 (section 4.8). For equivalent fluid pressure, it is suggested that Ψ should be taken as unity. Armer and Moore (1989) should be consulted for further information on the strip approach and, in particular, the advanced strip method, which is intended for the design of slabs supported as a whole, or in part, by columns.

REFERENCES

Hillerborg A. (1975) *Strip Method of Design*, Viewpoint.

Wood R.H. and Armer G.S.T. (1970) *The Strip Method of Designing Slabs*, Building Research Establishment Current Paper 39/70.

Timoshenko S. and Woinowsky-Krieger S. (1959) *Theory of Plates and Shells*, McGraw-Hill, New York.

Armer G.S.T. and Moore D.B. (eds) (1989) *Frame and Slab Structures* (chapter 7, L.L. Jones, Hillerborg's advanced strip method – a review and extensions), Butterworths, London.

Appendix F

COLUMN DESIGN CHARTS

F.1 INTRODUCTION

Column design charts 1 to 4 are given in Figures F.1 to F.4 for rectangular and circular columns with a single grade of concrete, $f_{ck} = 30$ N/mm² with d_1/h and $d_1/d = 0.1$, and two grades of steel, $f_{yk} = 400$ and 460 N/mm². The basic equations and computer listing are given in Chapter 6. A comprehensive set of design charts for columns will shortly become available, and enquiries should be made to the authors.

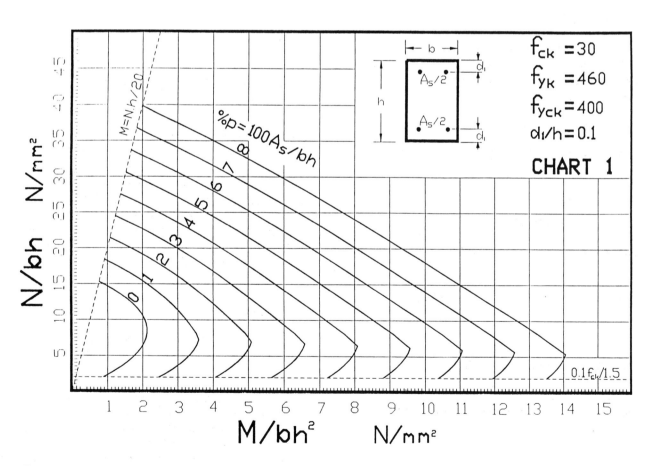

Figure F.1 Column design, chart 1.

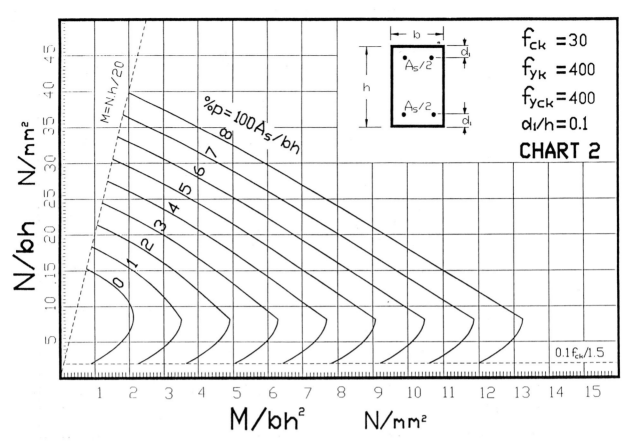

Figure F.2 Column design, chart 2.

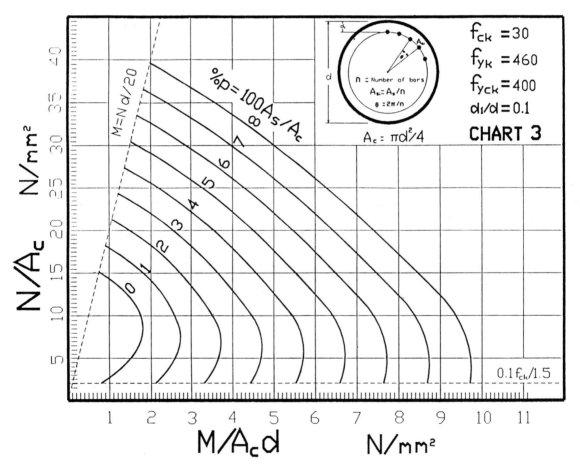

Figure F.3 Column design, chart 3.

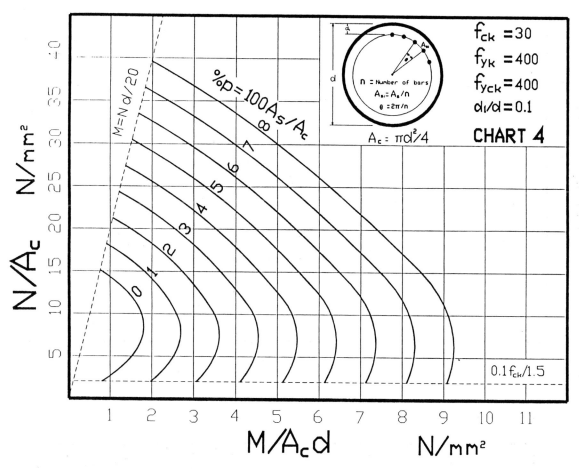

Figure F.4 Column design, chart 4.

F.2 USE OF DESIGN CHARTS

Rectangular column

Given:

$d_1/h = 0.1$ b = 400 mm $N = 3000$ kN
$f_{ck} = 30$ N/mm² h = 600 mm $M = 600$ kN m
$f_{yk} = 460$ N/mm² d = 540 mm

$N/bh = 3000 \times 10^3/400 \times 600 = 12.5$
$M/bh^2 = 600 \times 10^6/400 \times 600^2 = 4.17$

Thus from chart 1:

$100A_s/bh = 2.25$
$A_s = 0.0225 \times 400 \times 600 = 5400$ mm²

Circular column

Given:

$d_1/d = 0.1$ d = 600 mm
$f_{ck} = 30$ N/mm² $N = 3000$ kN
$f_{yk} = 460$ N/mm² $M = 600$ kN m
$N/A_c = 3000 \times 10^3/0.785 \times 600^2 = 10.62$
$M/A_cd = 600 \times 10^6/0.785 \times 600^3 = 3.54$

Thus from chart 3:

$100\,A_s/A_c = 2.0$
$A_s = 0.02 \times 0.785 \times 600^2 = 5652$ mm²

Appendix G

BEAM DESIGN CHARTS

G.1 INTRODUCTION

Beam design charts 5 and 6 are given in Figures G.1 and G.2 for singly reinforced beams with two grades of steel, $f_{yk} = 400$ and $460 \, \text{N/mm}^2$, and are derived from the parabolic rectangular stress–strain diagram for concrete given in section 4.2 of EC2. A simplified design chart based on the equivalent rectangular stress block is shown in Figure 4.3 (or B.2). A comparison between the two charts is given below.

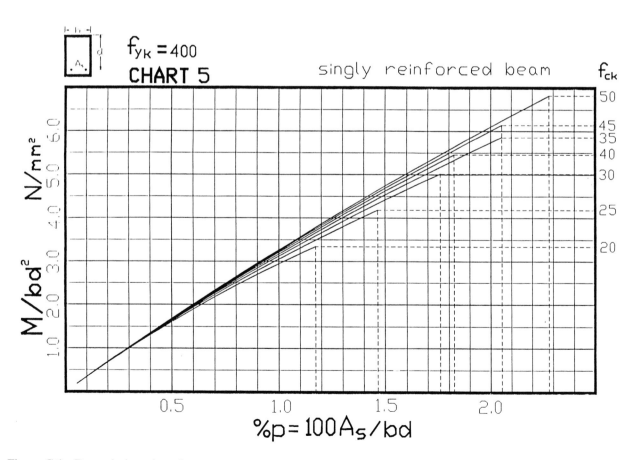

Figure G.1 Beam design, chart 5.

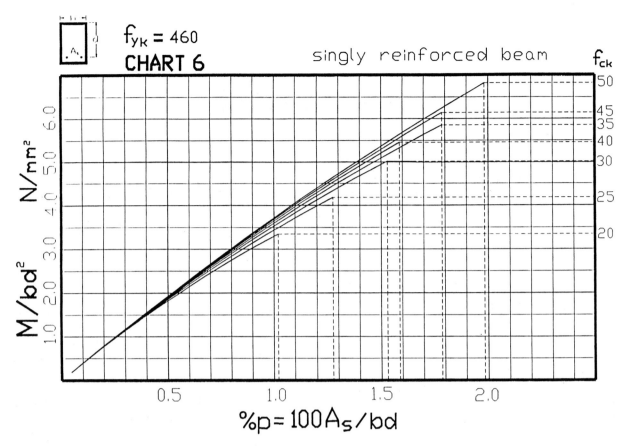

Figure G.2 Beam design, chart 6.

G.2 USE OF DESIGN CHARTS

Given:

  $b = 1000$ mm $f_{yk} = 460$ N/mm^2

  $d = 200$ mm $M = 100$ kN m

  $f_{ck} = 35$ N/mm^2

  $M/bd^2 = 100 \times 10^6/10^3 \times 200^2 = 2.5$

From Figure 4.3,

  $x/d = 0.14$

  $x = 0.14 \times 200 = 28$ mm

  $d - 0.4x = 188.8$ mm

Thus

  $A_s = 100 \times 10^6/(460/1.15) \times 188.8 = 1324$ mm^2

From chart 6, with $M/bd^2 = 2.5$

  $P = 100\ A_s/bd = 0.68$

Thus

  $A_s = 0.0068 \times 10^3 \times 200 = 1360$ mm^2

The difference in steel area is less than 3%, but there is less work involved in using charts 5 and 6 as the steel percentage is obtained directly. However, Figure 4.3 gives x/d values that are of use when applying plastic analysis or moment redistribution.

Appendix H

TYPICAL DETAILS FOR A FRAME DESIGNED FOR SEISMIC ACTIONS

H.1 DETAILING RULES

Detailing rules for beams, beam–column joints and columns are included in Chapters 4, 5 and 6. Broadly, the volume of confinement reinforcement required increases with increase in ductility level, whereas the percentage of longitudinal steel (in beams and columns) increases with reduction of ductility level. The following general points should be noted:

1. A hoop is defined as a closed stirrup with 135° bent-in ($10d_h$ long) hooks, where d_h is the diameter of the hoop reinforcement.
2. The spacing of the hoops (s_w) for beams, beam–column joints and columns should be in accordance with the rules set out in Chapters 4, 5 and 6.
3. The reinforcement ratios (maximum and minimum) should be in accordance with the values given in Chapters 4, 5 and 6.
4. At least one intermediate bar (between corner bars) should be provided at each column side for *all* ductility levels.
5. Single hoop confinement reinforcement should *not* be used in columns for ductility level 'H' structures.

ACI (1991) and Cheung *et al.* (1993) give further information on detailing for seismic actions.

Figures H.1 and H.2 show a typical arrangement of beam and column reinforcement in a reinforced concrete framework designed to ductility level 'M'. The reinforcement shown is based on the preliminary calculations set out in Chapter 8.

REFERENCES

ACI (1991) *Design of Beam–Column Joints for Seismic Resistance*, ACI SP-123.

Cheung P.C., Paulay T. and Park R. (1993) Behaviour of beam–column joints in seismically loaded reinforced concrete frames, *J. Inst. Struct. Eng.*, **71**, No. 8 (20 April).

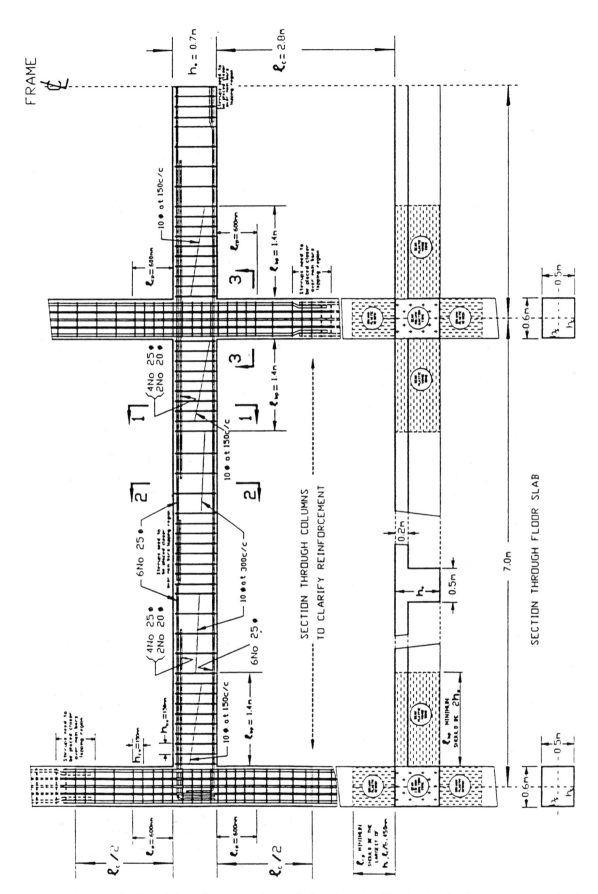

Figure H.1 Typical details for reinforced concrete frame designed to ductility level 'M': sheet 1, beam and column arrangement.

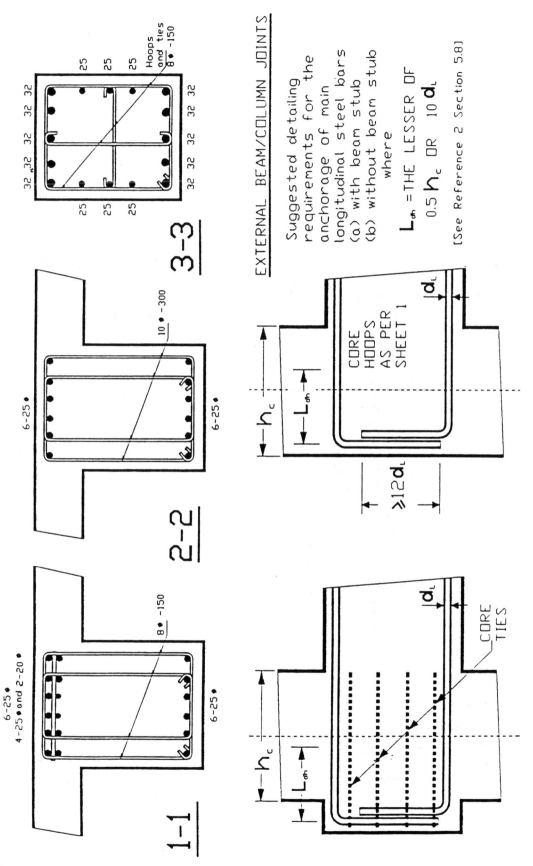

Figure H.2 Typical details for reinforced concrete frame designed to ductility level 'M': sheet 2, details of sections.

INDEX